EVIDENCE IN SCIENCE

EVIDENCE IN SCIENCE

A SIMPLE ACCOUNT OF THE PRINCIPLES OF SCIENCE FOR STUDENTS OF MEDICINE AND BIOLOGY

BY

KENNETH STONE, D.M. (Oxon.), M.R.C.P. (Lond.)

Emeritus Consultant Physician, The Arthur Stanley Institute for Rheumatic Diseases, The Middlesex Hospital; formerly Rose Research Fellow, St. Bartholomew's Hospital

BRISTOL: JOHN WRIGHT & SONS LTD.
1966

Distribution by Sole Agents:
United States of America: The Williams & Wilkins Company, Baltimore
Canada: The Macmillan Company of Canada Ltd., Toronto

PRINTED IN GREAT BRITAIN BY JOHN WRIGHT & SONS LTD.
AT THE STONEBRIDGE PRESS, BRISTOL

PREFACE

A VERY celebrated physician used to tell his students that the best textbook they could buy was Jevons's *Primer of Logic*. It was usually taken to be just a sample of a mordant sarcasm for which he was notorious, and which was always provoked by our more outrageous *non sequiturs*. I never knew of anyone who bought the book. But the advice was given seriously: he thought logic an indispensable study for all who are in complete ignorance, as most students are, of those simple principles of reasoning which are employed, consciously or unconsciously, in all scientific work.

Probably no better elementary exposition of Aristotelian logic has ever been written than Jevons's *Primer*. But Aristotle's logic was mainly concerned with the mode of inference called 'deduction', and this is but a part of scientific method. I doubt if it is what should have been chosen to recommend to a scientist as the best guide to straight thinking. What science students need is a simple exposition of the kind of thinking that guides the pursuit of scientific knowledge.

It is strange that the 'principles of science' is not a compulsory subject in any science student's curriculum. On these principles rests the art of investigating truth by means of evidence; and few will deny that, quite apart from its intrinsic interest, this is a highly useful art for those who undertake original research, and perhaps far more so for those who merely wish to read or listen and decide whether evidence has been found.

Since Locke expressed doubt that God had left it to Aristotle to make men rational it has often been objected that we can think correctly without knowing the rules, and that in any case reason has very little to do with discovery. Of course, some men can think without knowing the rules, or the rules could never have been discovered. Some men can write well, having made no study of the rules of writing. But most of us write badly, and they must be few who never reason incorrectly. A comment on the second objection will be found in the text.

In writing this book I have had in mind especially the needs and interests of students of medicine and other biological sciences. Most of them are so engrossed in their ever-increasing curriculum that they have not enough leisure to read an immoderately exhaustive book, as are most of the textbooks dealing with the principles of science. Nor do they wish to be involved in arguments about the more abstruse problems: I have tried to omit controversial topics, and I hope I have not put forward any views not commonly accepted by logicians. Almost all the illustrative examples are taken from the biological sciences, especially those contributing to medicine, and they are deliberately presented in rather more detail than is usual. Even if I fail to persuade anyone to pay attention to the correctness of his reasoning, I hope I may succeed in holding his interest.

I hope also that the book may have an interest for the general reader who wishes to inquire into the scope and nature of scientific knowledge and the possibility of understanding the world around us better than we do. I have tried to give a glimpse of how men in the past have tried to deepen their understanding, of how they failed, and of the new way of thought of those who, at the beginning of our scientific era, 'first broke the ice for us, and smoothed the first approach to the heights'.

It remains for me to thank very warmly Mr. T. E. Cowan, Statistician to the Middlesex Hospital, for reading in manuscript the chapters on 'Observation and Errors of Observation' and 'Mathematical Probability in Research'.

June, 1966 K. S.

CONTENTS

EVIDENCE IN SCIENCE

CHAPTER I

CREDULITY

THE roots of some beliefs must find very strange nourishment. I remember a man lying on my examination couch instructing me not to trouble about the left half of his body; it had been dead for years. He was irritated when I applied a stethoscope to the apex of his heart, and his irritation mounted to a flush of anger when I tried to disabuse him of the notion. I was young at the time and did not know the futility of it: you cannot destroy a belief like this.

Such odd beliefs are not as uncommon as you might think. I remember another man just as firmly convinced that he was inside out.

Psychiatrists, no doubt, can find what it is that nourishes these strange plants. You and I are content to call it 'lunacy'. The origin of the first was quite possibly a lesion in the posterior part of the right half of the brain. Loss of half the body image of this origin is well known: one patient, recorded by Brain,* dried only the right side of his body after a bath, but most people would agree that he was mentally disordered.

What then would we call our own readiness to believe that misfortune will follow if we walk under a ladder, that a political party will do what it says, or that the doctor's medicine will help? Superstition, credulity, or faith—not lunacy, we would say emphatically, for we can give you a reason for our belief, and a lunatic cannot.

Those whose bizarre beliefs come from some defect of the brain would, if challenged to justify them, simply reiterate: 'I know', whereas we can either give you the evidence for our belief or refer you to authority.

If we do ask ourselves why we believe something we usually find either: (1) Evidence that must convince all rational men; (2) Evidence that is clearly insufficient; (3) That we are trusting the word of someone who we believe could produce convincing evidence; or (4) That we are merely trusting someone's word. When the result of this introspection puts us in category 2 or 4 we are to be described as *credulous*.

When the belief concerns some fundamental concept like the uniformity of Nature, or a religious tenet, we say it springs not from credulity but from faith. 'Faith' in its broad sense means trust in the word of

* BRAIN, W. RUSSELL (1951), *Mind, Perception and Science*. Oxford: Blackwell.

someone whom we feel can be trusted. 'Credulity' is the endearing habit of trusting and believing everyone.

Credulous people may have nicer characters than the chronic doubters, but they have done far more harm than good in the world. Credulity has been the main source of false beliefs, but it may be said that they are not necessarily harmful. Do we not all, for instance, think that we are better men than we are? Perhaps this is an instance of good coming from a false belief, for the result is usually more hope and self-confidence than we would otherwise have. And there are many other occasions when a false belief brings peace of mind.

But when a conscious belief guides what we do it is desirable that it should be true, particularly if what we do is important. Although most of our activities are guided by habit or instinct there are some occupations in which nearly every act is the result of a conscious and reasoned decision. These decisions, if they are not to be disastrously wrong, must be based on true beliefs. The practice of medicine is one such occupation and there are many others.

Medicine is an art in which true beliefs are of supreme importance, and yet a volume would be needed to expound the credulity of physicians, up to the not very distant past, and the follies and cruelties born of their false beliefs. The quaint ideas of their patients have been harmless compared with their own fanatic credulity through the centuries of recorded history.

Hippocrates is revered as the father of medicine and his progeny have inherited in full measure his tendency to generalize from a few observations. It was one of the Alexandrian physicians, Erasistratus, about 300 B.C., who first said that the chief cause of disease was 'plethora', i.e., an excess of blood. His successors believed him and treated most diseases by bleeding—clearly the logical thing to do. Some 22 centuries later doctors came to understand that, so far from being a sovereign remedy, there was nothing to be said for it under any circumstances.

The long journey through the centuries to our scientific medicine has been enlivened by a succession of monuments to medical credulity—the successive editions of the *British Pharmacopœia.*

The pharmacopoeia of 1618 contained close on 2000 remedies. There were lozenges of dried vipers, foxes' lungs, powdered precious stones, oil of ants, and oil of wolves. There was the prototype of all subsequent 'shot-gun prescriptions' invented by King Mithridates of Pontus in the first century B.C. to be an antidote to all poisons. They called the many imitations 'mithridates' and used them for all diseases. The mithridate of our pharmacopoeia of 1618 had 230 ingredients.

Later seventeenth-century pharmacopoeias show the enormous faith that the physicians of those days had in the therapeutic efficacy of animal extracts. Medicaments were made of blood, fat, bile, viscera, bones, bone-marrow, claws, teeth, hoofs, and horns of all sorts of animals; of cocks'-combs, cuttlefish, fur, feathers, hair, human perspiration, saliva of a fasting man, spiders' webs, sponges, sea-shells, casts of snake's skin, scorpions, swallows' nests, woodlice, and bones from the skull of an executed criminal.*

* Garison, F. H. (1957), *History of Medicine.* Philadelphia: Saunders.

In the eighteenth-century pharmacopoeias most of this rubbish has gone, but in the editions edited by Sir Hans Sloane a mithridate is still there, and there are preparations of crabs' eyes, woodlice, pearls, and vipers.

Very few of the remedies to be found in the nineteenth-century pharmacopoeias are used today, and it is not only the strange medicines of bygone times that have been given up. In my lifetime many treatments which we thought important and effective have been shown to be completely inert and worthless.

Usually no great harm was done. It has been pointed out that, though many of the old mixtures had little or no direct influence on the course of the disease, they were indirectly beneficial by the power of suggestion. The real cruelties practised in the name of Hippocrates have come from superstition, which is credulity regarding the supernatural.

Common to all folk-lore is the notion that some supernatural agency is the cause of disease and death. For primitive man it was obvious that there was a spirit world. All these moving things he did not understand —leaves rustling, queer flickering things in rippling water, shadows, and the creatures of his dreams—what else could they indicate? He was surrounded by spirits. And his life was so full of baneful things that these spirits must be mostly malevolent; so much was not a difficult inference. Disease clearly was part of their evil work.

As men developed a moral sense disease was seen to be not only the work of malevolent demons; often it was a punishment inflicted by the gods on men for their sins. *Flagellum dei pro peccatis mundi* it was called by a seventeenth-century divine, a vigorous persecutor of witches; and there are still some people who believe it.

The thing to do was to frighten off the demons, often by a charm of sorts, or to propitiate them by sacrifice. 'To coerce the spiritual powers', wrote William James, 'or to square them and get them on our side, was, during enormous tracts of time, the one great object in our dealings with the natural world.'

The conviction that we can get the spiritual powers on our side, which brought about the miraculous cures in the ancient Greek temples, at the tombs of the Saints, or at the shrine of Lourdes, persists in modern faith-healing. Sometimes only good seems to come of it, but more often it brings only heartbreak.

But for sheer evil born of superstition no example can surpass the treatment of the mentally ill throughout the Christian era to the early nineteenth century.

In the Dark Ages mental illness, well known to be possession by an evil spirit, meant, as witchcraft did, that the victim had been collaborating with the Devil. Anyone who collaborated with the Devil had to be tortured and burned alive, so the Fathers of the Church devoutly believed. Among the many signs of such collaboration were some of the now well-recognized symptoms of mental disease such as hearing voices and seeing visions. Even the presence of patches of anaesthesia in the skin was proof; the Devil, it was well known to the priests, marked some of his collaborators in this way. Many a poor woman has been tortured and burned alive after they had prodded her with a bodkin and found these numb spots.

Many of the patients in our own mental hospitals would have been burned at the stake in the sixteenth century because of this false belief. One hundred thousand at the lowest estimate were killed like this.

The doctors of the seventeenth century were almost, but not quite, as credulous as the priests. They still believed that the mentally ill were possessed; an evil spirit inhabited the body of the victim. As an abode, therefore, it should be made highly disagreeable to the toughest spirit. In England 'houses of correction' were established by Act of Parliament passed in 1609 for the treatment of lunatics. Two types of lunacy were recognized: melancholia and madness. While melancholic persons might be best treated by being entertained with 'mirth and merry', general medical opinion was in favour of harsh treatment for the mad. They were kept naked, cold, and half-starved. Every year they were given two months of 'reducing treatment' consisting of repeated bleeding, emetics, and purgatives. If frost-bite, gangrene, and death had not put a merciful end to it, few remained violent, and those who did were put in heavy chains and beaten. For every doctor knew that the best treatment for madness was fear, and fear could be kept up by repeated beating and threats of it.

By and large it is best that our beliefs shall be true. And I have said nothing so far of the men who want knowledge for its own sake, without giving a thought to its use. To a pure scientist, a 'natural philosopher', the pursuit of truth is the breath of life. It is rare indeed for a scientist to be credulous inside his laboratory—outside, perhaps!

I doubt if there is any remedy for credulity; it is probably an innate compensation for a defective reasoning faculty, and you cannot change a human mind. But if more and more of those not born incurably credulous will strive to think scientifically, there will be fewer lies to tell those who are.

It may well be objected that even scientists accept most of their beliefs from someone else. And this is inevitable. For some assertions, outside our own special domain of thought, we might ask for the evidence to be explained to us; probably we would not understand it. We must either have faith in authority or be content to know only an infinitesimal part of what is known. We cannot check every fact or assess every argument. But there is this sharp difference between the special faith that a scientist has in the word of a known authority on his subject and the common trust of the credulous: the one is born of the knowledge that his authority is guided by the kind of reason that scientists in general regard as valid; the other has no need of reason, or has never heard of it.

What it is to think scientifically, and how far reason can take us in our pursuit of truth, is the subject matter of this book. It may be that we shall conclude with F. L. Lucas: 'I believe that the most reasonable minds are cautiously distrustful even of reason itself.'* Even so, we will readily concede that reason works better than credulity.

---

* LUCAS, F. L. (1960), *The Greatest Problem.* London: Cassell.

CHAPTER II

TRUTH AND FALSITY

DEMOCRITUS said: 'Nature has buried truth deep in the bottom of the sea.' He probably meant that knowledge is remarkably elusive; but it could be that he thought the human mind incapable of knowing anything with certainty. Is the pursuit of truth as hopeless as this?

Different people have different ideas of what the word 'true' means. I remember reading in the correspondence columns of a journal an indignant protest at an assertion by an eminent contributor that it was difficult to name any eternal truths. The writer could name plenty: 'Two and two are four', he suggested.

Many others have imagined 'two and two are four' to be an eternal truth. Thus the Dean of St. Paul's: 'I don't believe that two and two make four: I know it, and if I understand it I cannot conceive it to be false. . . . No one has ever agonized in prayer for faith to believe that $2 + 2 = 4$.' And G. K. Chesterton, when he wrote the bank clerks' chorus:—

> Up my lads, and lift the ledgers, sleep and ease are o'er.
> Hear the Stars of Morning shouting: 'Two and two are four.'
> Though the creeds and realm are reeling, though the sophists roar,
> Though we weep and pawn our watches, two and two are four.*

But 'two and two are four' is a nominal definition of 'four'. It states that whenever I wish to refer to 'two and two' I shall do so by the less awkward name 'four'. A *nominal definition* is a statement of our intention to use words or symbols in a certain way; it makes no assertion about the subject to which they refer. In the words of Whitehead and Russell: 'A definition is, strictly speaking, no part of the subject in which it occurs. For a definition is concerned wholly with the symbols, not with what they symbolize . . . it is not true or false.'†

The definitions of mathematics are of this kind. 'An equilateral triangle is a three-sided figure which has three equal sides' asserts no more than that we can economize words by using the shorter 'equilateral triangle' for the longer descriptive phrase. And, again, the definition we give to any name we have coined is clearly no more than a statement of verbal equivalence. 'A streptococcus is a spherical micro-organism which multiplies by division in one plane of space only' is not a proposition in bacteriology; it states merely that whenever I wish to refer to a micro-organism with these characters I shall do so by this name.

A nominal definition cannot be questioned, for it is not a kind of statement which can be either true or false; it is either correct or incorrect. It

---

* Quoted from the essay, 'Little Birds who won't sing', in *Tremendous Trifles*, by kind permission of Miss D. E. Collins.

† WHITEHEAD, A. N., and RUSSELL, B. (1910), *Principia Mathematica*. Cambridge: University Press.

is an expression of verbal equivalence we have agreed on, and that we have agreed on it is, of course, true or false; but truth or falsity is not applicable to the expression itself. The importance of this is that, whereas from general assertions that are true we may be able to infer other truths, from a convention about the use of words or symbols we can infer nothing. A nominal definition cannot be used as a premiss in reasoning.

A physicist may define 'force' as the product of mass and acceleration: $F = ma$. But he cannot infer from it the first law of motion, saying that if F is zero, a must be zero: in other words that 'every body will maintain its state of rest, or of uniform motion in a straight line, unless compelled by some external force to change that state'. If he did he would be extracting a truth about the physical world from a convention about the use of symbols.

He may say, and some will agree, that this definition is not nominal: it is a so-called *real definition*. If we analyse the concept that we think the name indicates, at first knowing it only vaguely, and express our analysis as a definition of the name, the procedure is quite different from that of declaring verbal equivalence. We are trying to make clearer the nature of the named concept and we are therefore asserting something about it. Such definitions are either true or false, and so may be used as premisses in reasoning. The physicist may say that $F = ma$ is the same kind of definition as 'force is defined as that which changes or tends to change the state of rest or uniform motion of a body', and that like this it says something about the physical thing which 'force' denotes.

But if we allow him this, he cannot then say, as some do: 'The first law of motion is "true by definition"', implying that, being a definition, it cannot be questioned. It is only nominal definitions that, being neither true nor false, cannot be questioned.

I shall assume that we all know what we mean by a 'fact': something existing or that has existed in the actual universe. I must skate lightly over the difficulty of defining 'exist'. A fact cannot be called 'true' or 'false'. An inkpot is a fact: to say an inkpot is true has no meaning. A sentence on this page is a fact, and it cannot be said to be true or false, but what can be so described is the *meaning* of certain sentences.

All that can be called true or false is a belief, or 'proposition', and this is expressed by the kind of sentence that affirms or denies something. Language has the remarkable property of making us aware of our own thoughts and of conveying thoughts from one mind to another. When we hear or read a sentence we can, more often than not, grasp the meaning that evoked it in the first place. It is this meaning, when the sentence affirms or denies something, that logicians call a 'proposition'. The sentence 'This light is red' is a series of marks on paper, or, if spoken, a series of sounds; the meaning that it conveys to the reader or listener is a 'proposition'.

Only propositions can be meaningfully called 'true' or 'false'. If there were no minds to formulate propositions and entertain beliefs nothing in the universe could be described as 'true'. It would contain only facts. If minds were destructible there could be no such thing as 'eternal truths'.

When there is a fact corresponding to a proposition the proposition is said to be true; when there is no such fact, false. Thus, if there is a fact corresponding to 'This light is red', the proposition is true; if it describes no fact it is false.

We can now take up the old Greek philosopher's gloomy thought: that perhaps truth will for ever elude us. Are there any propositions that are demonstrably true?

If we knew of one, what we should know would be the fact that makes it true. You may say, then, that we know an almost unlimited number of such facts, namely, facts that we can observe. Truths directly verifiable by observation are expressed by *singular propositions*, such as: 'There is an elephant on the road', or 'There is a rat in my room.' It may seem clear enough that all singular propositions can be known to be true or false.

Unfortunately, I might have delirium tremens. Those who study the subject called 'theory of knowledge' would say that, even if I have not, all I really know is the perception of a certain shape and patch of colour, from which I *infer* that I am seeing an elephant or a rat. There is always a big element of inference from the basic percept whenever we say that we see something.

The percept is all that we really know. And it follows that when we believe that we have observed some fact, no matter what, we may be mistaken. For that part of the belief that is arrived at by inference from the sense impressions we have had is quite often mistaken, as when driving at night we stop for an obstruction in the road and it proves to be the shadow of a bridge; or when I am awakened with a start in the night by an intruder breaking in a door downstairs, and it is a slate sliding off the roof.

Can we then never be sure of any observation? For the practical purposes of life of course we can. The scientific pursuit of truth proceeds from the known to the unknown; and the known starting-point is the fact of observation. Nevertheless, in theory we cannot be certain of any, and in practice we are often mistaken. Errors of observation beset the path of every research worker.

Moreover, these particular truths are not the kind in which scientists are interested, except as data for the truths they want to find. Science is about *general propositions*. By 'scientist' I do not mean 'technologist': I mean the men who used to be called 'natural philosophers', who are just trying to understand the universe around us, pursuing truth for its own sake and not for any use that it may have.

If we knew any general proposition to be true we should know some 'general fact' that made it true. The scientist's problem is: Are there any general facts that can be known, and how can we come to know them? We cannot know any general fact by observation. Observation could never demonstrate 'all dogs bark' or 'every animal that has lungs has two ventricles to its heart'. We could not observe all dogs, or every animal with lungs; and moreover, the fact described by a general proposition is not a fact of observation.

We could, of course, observe 'all dogs in this village bark'; but this is simply a short way of asserting a singular proposition about each member of a finite and observable collection. The general propositions which are the laws of a science do not express an enumeration of this sort. They are called 'propositions of unrestricted universality' because they assert or deny something not only of things that we have observed, but of things that we have not and never could observe. The fact they describe is a necessary association. 'All dogs bark' means that there is a necessary association between the class 'dog' and the class 'barking animal'. A necessary association is something not observable.

If we could deduce a general proposition from other propositions known to be true we could then be certain of its truth, for whatever is implied by a true proposition is itself true. But, as will be clear later, we can deduce a general proposition only from another, and this merely puts back the difficulty.

General knowledge, the kind of truth that seemed so well hidden to Democritus, is not to be had merely by observation or by deductive reasoning, or, it would seem, by any mode of inquiry other than that which has brought a qualified success during the last three or four centuries. And it is the kind of knowledge for which men have striven for at least two and a half thousand years.

CHAPTER III

KNOWLEDGE AND BELIEF

KNOWLEDGE is a category of true beliefs, namely, those beliefs shown to be true by evidence. As we are to inquire into the nature of this evidence it will be well to begin by removing a very common notion that strength of belief is an indication that what is believed is true, or likely to be true.

A feeling of absolute conviction that something is true gives some people a quiet confidence that it is so. But this is a common error. The medieval clerics were certain that all the stars of heaven revolved around an immovable earth. Until not much more than a century ago there was scarcely anyone who doubted for a moment that man and all living things had come to the earth by a special act of creation some 6000 years previously, and that all animal and plant kinds had remained precisely as they were in the beginning. They were quite certain of this, as certain as we are that there are no uncaused events; or as certain as the old Greeks were that everything that happened, happened for some purpose. The lunatic who believes that he is Julius Caesar is so strongly convinced of it that he becomes severely annoyed if you doubt his word.

For some people 'I know' expresses this complete conviction. 'I know that we survive death,' they might say, 'it is no use your arguing with me.' But this is a mistake. 'I know' is not the same as 'I firmly believe'. Nor is it the same as 'I have a true belief'. A savage, having squared the spirits, believed that he would recover from his illness and often did; when he did he had had a true belief, but not knowledge.

'Knowledge' refers not only to the belief but to the evidence for it. We can only say 'I know' when we mean 'I know evidence that gives this belief the highest credibility, that is certainty'.

Consider any single assertion, such as 'John is taller than Jane'. What do I really know if I believe this? All I can know is the evidence for its truth. After measurement I may say that I know 'John is taller than Jane'; what I know is the evidence of my measurement. Suppose I have never met either of them. If they are brother and sister, both healthy children, John aged 10 years and Jane 4 years, I may still say that I know 'John is taller than Jane'; but what I know is the evidence, namely, these facts and some simple logic. This evidence gives the assertion a high degree of credibility but not quite the same certainty. If again John and Jane are not related, their ages are 10 years and 8 years, and I have no information about their health, then I am far from certain, and would hesitate to say that John is even probably the taller.

The evidence for the truth of a proposition may give any degree of credibility from zero to maximum, which is certainty. What I am contending is that a feeling of conviction can never be taken to be part of this evidence.

Clearly we often have a strong feeling of conviction even when we can produce no evidence at all for the truth of what we believe. We find it difficult to say why we have it. Its origin, when we cannot account for it, is in the subconscious mind and therefore difficult to trace.

There is no doubt that one source is the wish to believe whatever it is. Subconsciously we want to believe that life has a purpose, that man is something more than an intelligent animal, that mind and body are distinct, and so forth. And the strength of a belief that fulfils a subconscious desire or dispels a subconscious fear is unshakeable by any argument.

I am not suggesting that what we wish to believe must be false, merely that a feeling of conviction of this origin cannot be taken as evidence for its truth. There is an infallible criterion by which we can recognize the conviction that stems from subconscious desires and fears. It is that when the belief is challenged we respond with an upsurge of emotion, emotion perhaps stirred by the fear of losing a belief that brings peace of mind. Whatever the reason, this is how people always react when a faith born in the subconscious is challenged. Mistrust a strongly held belief if you find that in defending it you lose your temper and abuse your opponent. Only trust it if you can defend it against criticism as calmly as you would the theorem of Pythagoras or the atomic theory.

We may be tempted to say of a feeling of certainty that it is only significant when brought by strong evidence; whenever we have it unaccountably we should ignore it. But ask any research worker whether it is not true that some ideas come with a complete conviction of their truth, others not. Time and again I have heard men say, 'This is a true bill', or, as my old chief used to put it, 'I've got the right pig by the ear this time'. Some of the most famous men in scientific research have recorded the feeling of complete certainty that came at the birth of a theory, later proved true. Henri Poincaré told how on several occasions concentrated work on a difficult mathematical problem had been quite fruitless; then came sudden illumination accompanied by a strangely strong feeling of conviction that he had hit on a method by which the problem could be solved. He had many of these experiences, always with the same inexplicable sense of certainty.

No pure scientist can neglect a feeling of conviction that he is on the right path. The strength of his belief is not the evidence he is looking for, but it may be a signpost to it.

CHAPTER IV

THE ORIGINS OF SCIENTIFIC METHOD

EVIDENCE for any belief consists of certain facts and certain principles of inference. The difficulties in the past have been to know which facts and which principles of inference.

To give knowledge, inference must proceed from the known to the unknown: the known is its raw material. But what could be taken as known?

The ancient Greeks conducted their inquiries into a problem by deductive reasoning from what seemed to them to be self-evident truths. This explains why they achieved so much in mathematics and so little in their attempts to understand the actual world. The only way in which reasoning can lead to knowledge that is certain is by beginning with assertions about whose truth there can be no doubt. In mathematics there are assertions that, if we are to think at all, we must admit to be self-evident truths; and from them deductive reasoning may safely begin. Euclid proved all the theorems of plane geometry by rigorous deductive reasoning from a few axioms such as: 'Things equal to the same thing are equal to one another.'

But there is no general statement about the external world that is a self-evident truth. The old Greeks thought there were plenty. They thought it self-evident that everything that happened on earth had some purpose. They thought it a self-evident truth that planets move in circles, for heavenly bodies, being divine and perfect, must move in the perfect figure, which is the circle. From this it was quite safe to begin astronomical inquiries. Plato and his followers thought that reasoning could safely start from any general truth given by 'intuition', which they believed to be an awakening of knowledge that we bring into the world with us.

'It is easy to distinguish those who argue from facts and those who argue from notions', said Aristotle, with a gentle contempt for the thought of his time. There is some reason to believe that by 'facts' he meant those particular truths that can be verified quite simply by observation; although in practice he obviously counts as facts a number of 'notions' that seem to him to be axiomatic. At least he was insisting that we argue from the known; and this is part of the foundation of modern scientific method.

Aristotle never put scientific method as we know it into practice, but in biology he was the first man to make detailed observations. Read an English translation of Περὶ Ζῴων (*Historia Animalium*) and you will meet a man absorbed by the mystery of the multitude of living things in our world. No one knew any more about them than the local fishermen or huntsmen, and there were no books on biology. So he wrote down all that he saw, so that his pupils and successors should at least have a body of recorded fact to help them.

This man was the world's first and greatest biologist. Twenty-three centuries ago, off the coast of an Aegean island or on some river in Macedon, this wealthy, elegant, and fastidious scholar would lie prone in a little boat, hour after hour, intently gazing through the clear waters at the strange creatures he saw, so that he and others could 'argue from facts'. His was the kind of mind fascinated by fundamental problems and excelling in abstract thought. One does not expect such a man to be human enough to like the wild creatures of the earth, or patient enough to observe in minute detail their form and structure and habits, as he did.

But we have only to look at Aristotle's physics to see how hard it is to challenge the fundamental assumptions of our fellow men. It is easy for us to show derisively the unverifiable 'notions' from which Aristotle himself argued if we conveniently forget that at times we do much the same. It was to Aristotle and all his contemporaries an assumption that seemed a self-evident truth that the natural state of a terrestrial body is one of rest, and that, although they all have a natural tendency to fall towards the centre of the earth, what is unnatural and has to be explained is all other motion. Why is this, on the face of it, a sillier assumption than that of the law of inertia, that a body in motion would never of itself come to rest, and that what has to be explained is what changes or stops its movement? One is not sillier than the other: the difference is that one works and the other does not.

Aristotle's assumption that a body keeps moving only so long as a mover is in contact with it soon got him into difficulties from which only a fanciful explanation could extricate him. An arrow when it leaves contact with the bow-string does not promptly fall: why not? Aristotle's explanation was that displaced air rushed round to the back and pushed it on. The earth, of course, is at the centre of the universe; the moon, the planets, the sun, and the fixed stars are clearly fixed to transparent impenetrable concentric spheres which roll unceasingly around the earth. What moves them? Obviously there are movers of some sort, and what more likely than intelligent spirits? But these assumptions were facts to Aristotle. His views about the universe were arrived at by valid reasoning, starting, he thought, with what could be taken as known.

There were other dissentient voices in ancient Greece, urging men to base their opinions on observation: scientific medicine began about 400 B.C. with the clinical observations of the Hippocratic physicians. In pre-Hippocratic times disease was thought to be a punishment inflicted by a god or by the magical powers of a human enemy. A physician could show the sufferer how, by suitable incantations, he could placate the god or make the effects of a human curse reversible.

Hippocrates (460–370 B.C.) insisted on exact clinical notes. His followers were faithful to his teaching for about 600 years; during these centuries many unknown authors contributed to the books we know as the 'Hippocratic Collection', which contain the first recorded clinical observations. Everything is noted: 'The patient's habits, regimen, and pursuits; his conversation, manners, taciturnity, thoughts, sleep or absence of sleep, and sometimes his dreams . . . his picking and scratching, his tears . . . urine, sputa, vomitings . . . sweat, coldness, rigor, cough, sneezing, hiccup . . .'

They saw that they were presented with the same group of symptoms and signs in one patient after another, and they made hundreds of rough-and-ready generalizations: inductions that were the first groping and tentative efforts with which the science of medicine began. The aphorisms of the Hippocratic collection record some of them: 'Eunuchs do not take the gout, nor become bald'; 'Spasm supervening on a wound is fatal'; 'Stupor or delirium from a blow on the head is bad'; 'Sneezing occurring to a woman affected with hysterics, and in difficult labour, is a good symptom'; and so on.*

The physicians of ancient Greece were observant people, with the same intelligent vision as the poets and sculptors. But they could not progress beyond the stage of observation: for 600 years they could do no more than observe and make their simple inductive generalizations.

Aristotle died in 322 B.C., and the last of the biological investigators of antiquity, the physician Galen, in A.D. 200. During these five centuries there had been a little progress in the observation of living things, but very little. And with the gradual decline of the western half of the Roman Empire the long bleak centuries of medieval Europe began. During the Dark Ages that intervened between Greek civilization and our own the literary masterpieces of the fifth and fourth centuries B.C., including the biological works of Aristotle, were preserved in two regions of the world: the Mohammedan Empire and Byzantium, Asia Minor and Syria. Many Greek books were translated into Arabic, and scholars in western Europe began to discover these Arabic translations in the twelfth century. And so Greek literature first became known through Latin translations of Arabic translations of the original Greek. Aristotle's biological books were translated from the Arabic in 1217.

But the curiosity of the ancient Greeks was not reborn until the fourteenth and fifteenth centuries, when European scholars discovered that Greek was still spoken and read in the Near East, and that the old Greek manuscripts were still available. The Byzantine people had preserved Greek culture and the old masterpieces in the original Greek, during the centuries when western Europe relapsed into semi-barbarism. The search for these old writings went on, at enormous trouble and expense, until by the mid-fifteenth century nearly all had been found.

It was taught throughout the Middle Ages that we have two sources of knowledge. 'Two angels guide the path of man': Reason and Revelation. Men have always been immensely proud of their reasoning faculty, and it was far from being neglected. For St. Augustine reason was God's gift to man which made him superior to the brute creation.

But still no one argued from what we should call 'facts'. Medieval reasoning was purely deductive. Aristotle's logical writings were used as a textbook; and Aristotle's logic does not extend much beyond an analysis of deduction. The thinking they did was to reason deductively from doctrines that the Church decreed must be accepted. The 'known' from which their deductions began was theological dogma: Aristotle's physics was what the Church said had to be believed about the external world,

---

* Quoted from *The Genuine Works of Hippocrates* (1849) (trans. ADAMS, F.). London: Sydenham Society.

and his philosophy was made the basis of Catholic doctrine by Thomas Aquinas in the thirteenth century.

The conclusions of the medieval thinkers were true *if* the premisses from which they started were true: for this condition attaches to all deductive conclusions.

'I predict', said John Ray, 'that our descendants . . . will be tempted to pity our ignorance and to wonder that truths easy and manifest were for so long hidden, and were so esteemed by us, unless they are generous enough to remember that we broke the ice for them, and smoothed the first approach to the heights.'

The men responsible for the scientific revolution of the seventeenth century, and the Renaissance humanists before them, broke the ice for us, not by their discovery of truths easy and manifest, but by their discovery of the new mode of inquiry which we call the 'scientific', or the 'experimental', method.

The scientific revolution was not a triumph of reason over revelation. There was no conflict, and belief in the two sources of knowledge remained largely unshaken. But now the two angels no longer co-operate. Revelation alone guides man in his quest for knowledge of morality and the supernatural. Reason, a reason of greater scope and power, will alone guide him in his search for knowledge about the external world.

Revelation had in fact said comparatively little about the world around us; the medieval Church had severely discouraged inquiries concerning it. What Aristotle had said about it was quite enough. The things of this world were scarcely worthy of notice; only the world beyond the grave had any importance.

This dreadful doctrine was destroyed in the main by men who were devout Christians. Our countryman, John Ray (1627–1705), the most celebrated botanist of the seventeenth century, was one of them. For him, all animals, all plants, and everything else in the world are manifestations of the mind of God: we are intended to delight in the beauty of created things. 'There are those', he wrote, 'who condemn the study of experimental philosophy as a mere inquisitiveness, and denounce the passion for knowledge as a pursuit unpleasing to God, and so quench the zeal of the philosopher. As if Almighty God were jealous of the knowledge of men!' His work had enormous influence in freeing scientific inquiry from the Church's condemnation.

To destroy the foundations of all supposed knowledge, to doubt everything and start again, cannot have been easy or painless. 'Scientific revolution' suggests a short, sharp, explosive transition: it was long, slow, tentative, and hesitant. Its protagonists had their medieval precursors. Aristotle's physics began to be questioned as early as the fourteenth century, and by the end of the sixteenth his cosy universe of a central earth and celestial spheres for ever rolled around it by intelligent movers had completely crumbled. The heavens were not made of changeless incorruptible stuff as he had said: for in 1572 a new star appeared, shone very brightly during the following year, and disappeared in 1574. And in 1577 a new comet cut through Aristotle's impenetrable crystal spheres.

It was clear that, as the old Greeks had insisted, men must make their own fresh astronomical observations. New and very accurate observations of planetary motions were made by Tycho Brahe (1546–1601): a splendid collection of data which formed the foundation for Kepler's mathematical treatment.

Naturalists of the sixteenth century, like the physicists, saw that they too must look for themselves. Clear, precise, and accurate descriptions of what was observed now began to be recorded. Printing had been introduced in the mid-fifteenth century, and the steadily increasing body of fact could reach a comparatively wide audience. Leonard Fuchs (1501–66), commemorated in the fuchsias, produced a lovely herbal, with descriptions of plants based on his own observations, beautifully and accurately illustrated. The Belgian, Andreas Vesalius (1514–64), when professor of anatomy at Padua, completed *De Humani Corporis Fabrica*, with its almost incredibly beautiful anatomical drawings.

The change from the medieval to the modern outlook was slow in England. When John Ray went up to Cambridge in 1644 he found no new zestful spirit of inquiry among his teachers; they were still occupying themselves with the classics and metaphysical puzzles. He could find no one in the whole university to teach him botany or to help in any way with its study. He began to teach himself by collecting and describing the plants growing around Cambridge. By the end of his life he had minutely studied and described every known species of plant from Britain, Europe, and Eastern lands, in all some 19,000. Collecting plants, in order to describe the apparently endless distinct kinds, had for some time been pursued with immense zest on the continent of Europe.

Man's sole guiding angel now was his reason, without the help of theological dogma and the teaching of ancient philosophers inextricably woven with it. In the seventeenth century especially did men think it a glorious thing to be rational. They believed that nothing could be beyond the power of the human mind to understand; no problem so obscure that it could not be illumined by reason. But the reason they knew was deductive inference. It had strict rigorous validity: a deductive conclusion could not be doubted if its premisses were true. Aristotle had brilliantly shown why it had this unassailable validity.

But you could not reason deductively from a hard brute fact, as you could, say, from Aristotle's generalizations about the physical universe. If you had to start your search for general knowledge from facts of observation, inference had to take some other route. You had to adopt a different way of thought whose validity was not (and still is not) completely obvious.

By the end of the sixteenth century collecting facts had become an obsession: only very gradually did it dawn on men that understanding is not illumined by mere passive observation without the guide of any theory at all. It was seen that what is important is the relation of fact and theory, and that observations must therefore be experimental. A theory usually implies certain consequences; an observation deliberately made to see if these consequences are facts is what is meant by an experiment. In popular speech 'experiment' is often wrongly restricted to an observation made under artificial conditions.

Galileo Galilei (1564–1642) was among the first to turn to the experimental method, making it the basis of his studies in physics. *Discourses on Two New Sciences*, published in 1638, contain his most important scientific work. To illustrate the method, take the following investigation—one of many by which he founded the modern science of dynamics.

Galileo assumed that everyone had been wrong in thinking that a moving body would come to rest, even if there were no resistance, unless something kept it moving. He supposed the reverse: that a moving body keeps moving with the same velocity and direction unless some force acts on it.

There were several consequences of this supposition that could be tested by experiment: compared, that is, with what is actually found by observation. It is a consequence that a falling body falls with constant acceleration.

It can be shown by simple mathematical reasoning that the distance s covered by a body falling with constant acceleration for a time t, starting from rest and attaining a velocity v, is

$$s=\tfrac{1}{2}vt. \qquad (1)$$

A second supposition was that the velocity increases with the time: if g is a constant

$$v=gt. \qquad (2)$$

It follows from (1) and (2) that

$$s=\tfrac{1}{2}gt^2, \qquad (3)$$

i.e., the distance fallen varies with the square of the time.

The next step was to see if this consequence accorded with fact. But a freely falling body moved too quickly for accurate measurement. Galileo solved the difficulty by first showing that a body rolling down an inclined plane acquires the velocity it would have reached if it had fallen through the same vertical height. His experiment was made with a board 12 yards long, which could be inclined at various angles; cut in the whole of its length was a very smooth groove down which a polished brass ball could be rolled. Relative times were measured by a water clock: through a small opening in a large vessel water could be made to flow during the time to be measured into a small receptacle in which it could be weighed. The relative time intervals were given by the relative weights. Repeated observations showed that, with a fair approximation, the distance traversed by the ball from rest varied as the square of the time. To give one observation: the time to traverse the first quarter of the board was one-half the time to run its whole length.

The seventeenth-century revolution in thought, which made possible our modern world, was the discovery that the best way to find out if a theory is true is to see if its consequences are facts. The experimental method has increased in complexity but its simple principle remains the same.

CHAPTER V

ARISTOTLE'S THEORY OF DEDUCTION

INFERENCE is a process of thought which extracts the truth of a proposition from that of others: we assert that a proposition is true because we know that others are true. The peculiarity of *deductive inference* is that the propositions with which we start, called *premisses* ('proposition put first' —Latin *prae*=before; *mitto*=I send or put), *imply* the conclusion. If we know the premisses to be true the truth of the conclusion cannot be doubted. The principle of deduction is that what is implied by a true proposition is true.

Deduction is the reasoning of mathematics, the only purely deductive science. It can give us certain knowledge if we can start with anything certain. Look once again at Euclid's axioms. There are 12 of them, assertions of a kind that we are content to call 'self-evident': 'If equals be added to equals the wholes are equal', and so on. We cannot doubt them. Now look at the end of a chain of deductions from some of the axioms, say at the theorem of Pythagoras. The theorem is implied by the axioms, for nothing else has been brought in to contribute to its proof: it is therefore certain if the axioms are certain. And as we cannot doubt this we cannot doubt the truth of the theorem. Our knowledge of its truth is as near certainty as the human mind is capable of approaching.

Deduction can, theoretically, give certain knowledge; but can it give new knowledge, not already stated by the assertions with which it begins? Strictly, no—although it gives knowledge new to us. It must be true that in some manner the conclusion of a deductive inference is contained in the premisses. The theorem of Pythagoras must, however hard it may be to see how, be contained in the self-evident axioms from which Euclid starts his reasoning. The conclusion cannot possibly contain anything not in some latent fashion contained in the axioms, for nothing else has been introduced into the chain of inferences. But we should not be aware of the conclusion without the reasoning. Deduction *reveals* a truth contained in the premisses. It is an exposure of truths already there, and not a discovery of new knowledge going beyond anything already given.

The problem is, how can it be known when premisses imply a conclusion, and when they do not?

Aristotle (384–322 B.C.) was the first man to give his mind to this question.

Athens at that time had been the centre of learning in the Mediterranean world for more than a century, attracting students as Paris or Oxford drew them in the later Middle Ages. The average Greeks of the fourth century were men of a level of intelligence never surpassed before or since. Scholars are agreed that they were far more intelligent than present-day Englishmen. They had ample leisure and a thirst for knowledge. Athenian men loved to foregather in the colonnades that ringed the market-place, where they could be sure of shade from the Mediterranean sun, and where there could be heard a babble of voices skilfully arguing political, or ethical, or any other debatable problems.

Aristotle was the first among them to see a certain sameness in the reasoning process, whatever subject was being discussed. One of his fundamental notions is that, in considering whether an instance of deductive reasoning is valid, it does not matter what we are talking about. It is enough to consider the form of the argument.

The idea is exceedingly simple and illuminating.

Aristotle taught that every conclusion is deduced from two premisses, and that every valid deduction can be stated in the form of:—

All men are mortal,
Socrates is a man: therefore
Socrates is mortal.

He saw clearly that if an inference has this form it is valid, i.e., the premisses imply the conclusion.

The notion of the *form of an inference* is not difficult to understand. Change the subject and the predicate in the three propositions, and, though you will get a different inference, something obviously stays the same: this is the form of the inference. It would be admitted that:—

All tyrants deserve death,
Caesar was a tyrant: therefore
Caesar deserved death

has the same form as the argument deducing the mortality of Socrates.

The form of this argument is called the *syllogism*. A syllogism is a triad of propositions, of which two premisses jointly imply the conclusion.

Aristotle would have said that the argument:—

All men are mortal,
All Tibetans are men: therefore
All Tibetans are mortal

has the same form. Actually it is slightly different, for Socrates and Caesar are individuals, Tibetans a class. But it is covered by the same rule.

Aristotle believed that all propositions are of the 'subject–predicate' form: whatever it may be that we assert, he said, we are saying that some attribute qualifies, or 'is predicated of' some subject, as in 'all men (S) are mortal (P)'. Thus all propositions can be stated in the form: S is P.

It will be clear that the form of all the above syllogisms can be stated symbolically:—

All M is P,
All S is M: therefore
All S is P

where M stands for 'middle term', i.e., the term with which both subject and predicate of the conclusion have been compared in the premisses.

If the premisses are true the conclusion must be admitted to be true: for the reasoning is an instance of the rule that whatever is affirmed of a whole class (P of M) can be affirmed of that which is contained in the class (S); or, in the corresponding negative syllogism, whatever is denied of a whole class can be denied of that which is contained in the class.

This rule was thought to be the universal principle of all reasoning. In the Middle Ages it was named the *dictum de omni et nullo*=statement concerning all and none. It is usually expressed more precisely: *Whatever is predicated, whether affirmed or denied, of all the members of a class may be predicated in like manner*: (1) *Of any individual member*; (2) *Of every member of any class included in it.*

The rule decides whether a syllogism is valid. It is not always easy to see whether a syllogism is an instance of it; but this can usually be settled quite clearly by the use of a diagram in which circles represent classes.

Thus the class of Tibetans included in the class of men, and the class of men included in the class of mortal beings, are shown diagrammatically below. It can be seen that the class of Tibetans must be included in the class of mortal beings, and so the reasoning must be valid.

All M is P: All men are mortal,
All S is M: All Tibetans are men: therefore
All S is P: All Tibetans are mortal.

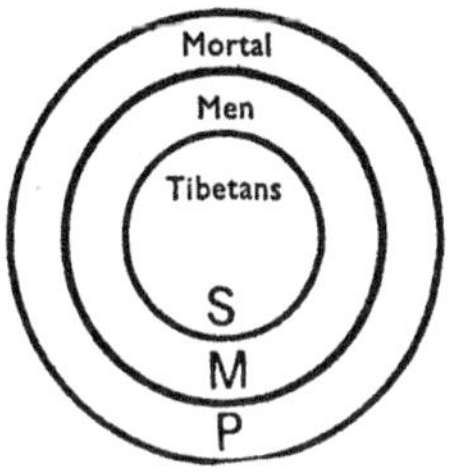

Consider another syllogism in which the constituent propositions are not all universal affirmative propositions, and the conclusion is negative:—

No M is P: Nothing that shows evidence of design has been produced by chance;
All S is M: All animals show evidence of design: therefore
No S is P: No animal was produced by chance.

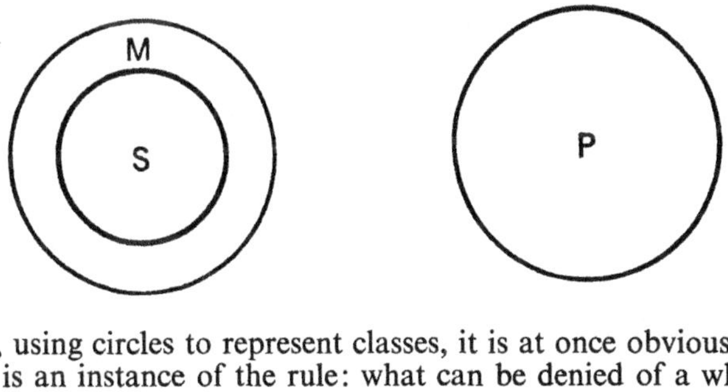

Again, using circles to represent classes, it is at once obvious that this syllogism is an instance of the rule: what can be denied of a whole class (P of M) can be denied of that which is contained in the class (S).

We may consider another of these slight variations in syllogistic form —one with a change in the position of the middle term in the premisses:—

No **M** is **P**: No one whose health is undermined by self-sacrifice is a source of happiness to others;
All **M** is **S**: All whose health is undermined by self-sacrifice are altruists: therefore
Some **S** is not **P**: Some altruists are not a source of happiness to others.

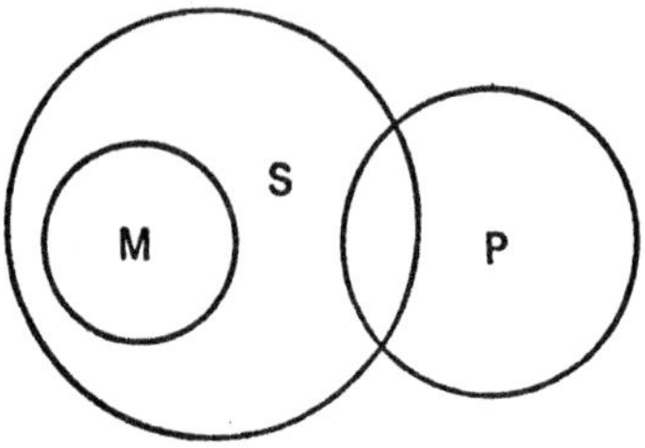

Again, a diagram makes it clear that what we are saying is: P can be denied of M, and therefore of what is contained in M, namely, some S. The syllogism is in accordance with the *dictum* and therefore valid.

Many triads of propositions appear at first sight as if they might be valid syllogisms. Perhaps the reader will suspect that the following are not valid, but he may not know why:—

White is a colour,
Black is a colour: therefore
Black is white.

All quadrupeds are animals,
A bird is not a quadruped: therefore
A bird is not an animal.

By using circles for classes he will easily see that they are not instances of the *dictum*. But it is not always so easy, and simple rules of validity have therefore been drawn up. Any syllogism that violates any rule is faulty: all that do not are valid. The *rules of the syllogism* are these:—

1. *Every syllogism has three and only three terms*: The *major* term, which is the predicate of the conclusion; the *minor* term, the subject of the conclusion; and the *middle* term, which does not occur in the conclusion.

2. *Every syllogism has three and only three propositions*: The major premiss, containing the major term; the minor premiss, containing the minor term; and the conclusion.

3. *The middle term must be distributed once at least in the premisses*: When we say something of the whole of a term, the term is said to be 'distributed'; when of only a part, 'undistributed'. The first of the spurious syllogisms given above is an example of *the fallacy of the undistributed middle term*: 'colour' being undistributed in both premisses. A diagram shows at once that we are comparing 'white' with one part of the middle term, 'black' with another. The same fallacy occurs when the middle term is ambiguous:—

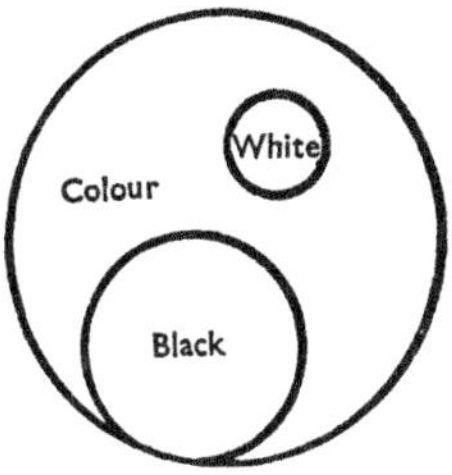

4. *No term must be distributed in the conclusion which was not distributed in the premisses*: Neglect of this rule is called *illicit process* of the major or minor term. It is invalid because we conclude something about the whole of the term, whereas in the premiss we said something only of a part of it. The second spurious syllogism above is an *illicit major*: the conclusion distributes its predicate, and in the major premiss this is undistributed.

5. *From two negative premisses you can infer nothing.*

6. *If one premiss is negative, the conclusion must be negative.*

At the end of his work, which appeared about 330 B.C., Aristotle wrote:—

> We have had no works of predecessors to assist us in this attempt to construct a science of reasoning; our own labours have done it all. If, therefore, the work appears to you not too inferior to the works of other sciences which have been formed with the assistance of successive labourers in the same department, you will show some indulgence for the imperfections of our work, and some gratitude for the discoveries it contains.

Aristotle's logical writings were translated into Latin in the fifth century, and with scarcely any addition of importance they were the substance of our textbooks of logic until about 100 years ago. For over 2000 years men believed that Aristotle had not only founded the science of logic but had completed it. As comparatively recently as 1836 Archbishop Whately wrote: 'The logical system is one of those few theories which have been begun and perfected by the same individual. The history of its discovery . . . properly begins and ends with Aristotle.'

CHAPTER VI

THE MODERN THEORY OF DEDUCTION

THE science of logic stagnated for about 2200 years. The syllogism is not the only form of deduction nor the most important. Mathematicians do not reason in syllogisms. In everyday life other deductive forms are much more common. Commenting on an expression of regret by the B.B.C. for the transmission of one of their programmes, someone asserted that this was not because of its bad taste. Asked how he knew, he said: 'I know by implication.' He had argued: If it is a principle to apologize for bad taste, then the B.B.C. would apologize ten times a day, and they do not: therefore it is not a principle to apologize for bad taste. Whether we agree with the 'if–then' sentence or not we have to admit that this is valid deduction, and is not a syllogism. Let p stand for any proposition, q for any other; if we can assert that p implies q and that q is false, no one can deny the conclusion that p is false. Put any propositions you like for p and q, and the inference is still valid deduction. We all reason like this many times a day; anyone can think of innumerable instances.

The traditional logicians managed, with great ingenuity, to twist most deductive reasoning into the form of a syllogism. They were defeated by 'a horse is an animal, therefore the head of a horse is the head of an animal'. This is valid deduction, but not a syllogism, and it cannot even be tortured into the form of one.

The limitations of Aristotelian logic were seen about 100 years ago; but modern deductive logic really dates from the publication in 1910 of *Principia Mathematica* by Bertrand Russell and A. N. Whitehead. One of the greatest intellectual achievements of all time, its theme is that mathematics is an extension of logic. Here I shall mention only the concept which is shown to underlie the validation of deductive inference.

The salient question which Aristotle tried to answer was: How can we know when the premisses imply the conclusion?

To understand the modern answer it is necessary first to introduce the concept of a *formal molecular proposition.* Some indicative sentences are composed of two or more sentences in such a way that the whole has a single meaning. Suppose I stand at a traffic crossing and say: 'If this light is red then that is green': although there are two constituent sentences I have not asserted two propositions, but one proposition. The whole has a single meaning which must be either true or false; therefore it is one proposition. We call *atomic* a proposition not composed of constituent propositions, for example: 'This light is red' is an atomic proposition. With the same chemical analogy we call a proposition *molecular* when it is composed of two or more atomic propositions joined together by certain relations.

The *form* of a molecular proposition can be expressed by letting p, q, and r stand for constituent propositions. Thus, the form of the argument in the first paragraph of this chapter is 'p implies q and q is false, implies p is false'. It can be, and usually is, expressed wholly in symbols. This symbolic statement is a 'formal molecular proposition'. When propositions are substituted for p and q it becomes a *material proposition.*

The answer to Aristotle's problem—the principle which guarantees that premisses imply a conclusion—is that the form of every valid deductive inference must be one of a group of formal molecular propositions with the peculiar characteristic that they cannot be false.

One has just been asserted: 'p implies q and q is false, implies p is false'. This proposition is always true, whatever may be the truth-value of p and q; p and q may each be either true or false, it does not matter. The whole molecular proposition is true whatever they are.

Here is another: 'p or q is true and q is false, implies p is true'. And a third: 'p implies q and q implies r, implies p implies r'. There are many scores of these logical tautologies, any one an implication that is true independently of the truth-value of the constituent propositions. Take any one, and make a material proposition of the same form, simply by putting some material proposition in place of p, q, and r. Then the material proposition is also an implication that must be true. The obvious question is: How do we know that these formal propositions are true, irrespective of the truth-value of their constituent propositions?

Euclid starts his *Elements* with 3 postulates and 12 axioms, which we are asked to accept without proof. They appear self-evident and we accept them. All the theorems of his plane geometry are then deduced, step by step, from them and from theorems already proved in the same way. The whole forms a deductive system in which every theorem is deductively connected with all the others, and all have the same certainty as the basic postulates and axioms from which they have been deduced.

Principia Mathematica unfolds a similar deductive system in logic. To demonstrate a chain of deductive inferences it is necessary to start with certain primitive propositions that we can agree to accept without proof. Unless we are prepared to accept some proposition, p, unproved, though we can demonstrate a chain of implications, we shall never be able to break it and proceed to an inference. If p implies q, but p cannot be asserted, we are left with a useless implication. We may go on and show that q implies r; and again that r implies s; but however long we go on all we can get is a chain of implications. But if we begin by admitting that a proposition, p, is so self-evident that we can accept it without proof, then, if p implies q we can assert q. And if we can then show that q implies r, we can assert the further theorem r, and so on.

Principia Mathematica starts its deductive system with axioms that we could not refuse to accept: primitive propositions of the level of simplicity of 'p or q implies q or p'. From these axioms all the formal truths validating deductive inference, and very many more similar theorems, are rigorously proved, step by step, using in the demonstration only the axioms themselves or some theorem already demonstrated.

In the preceding chapter, and so far in this, we have been concerned solely with the principle by which we may know when premisses imply a conclusion. But inference goes beyond the recognition of a true implication: this by itself is a fairly useless piece of information. What does it profit us to know that Caesar deserved death *if* all tyrants deserve death? We cannot proceed to an inference until we can assert the condition under which the conclusion is true; *p* implies *q* generates an inference if we know that *p* is true: it is the unconditional assertion of *q* that constitutes *inference.*

Reverting to material propositions and arguments, the logical rule tells us if we have a true implication, but only empirical knowledge can tell us if we can assert the antecedent and so proceed to an inference. Consider, as an example, William Harvey's exposition of his proof that the blood circulates. He reasoned: 'If either there is an exit for the blood, or the arteries must become so distended by incessant injections of blood that they burst, and the arteries do not burst, then there is an exit for the blood.' This is a true implication because its framework is the logical proposition '*p* or *q* and *q* is false, implies *p* is true'. Harvey's observations showed incessant injections of blood into the aorta: the left ventricle usually delivering, he calculated, about 15 litres of blood each hour. But the arteries did not burst. He could therefore assert the first part of the implication, and conclude that there is an exit for the blood.

He could not see an exit: his lens was not strong enough to see the blood passing from the arterioles to the venules. But he could be as certain that it did pass as he was of the truth of his premisses. The capillaries were first seen about half a century later.

Harvey wrote of this deduction: 'Now this, my conclusion, is true and necessary if my premisses be true; but that these are either true or false, our senses must inform us, not our reason—ocular inspection, not any process of the mind.' Deductive inference could not be described more succinctly.

It may be helpful to summarize the formal propositions that commonly validate deductive inference, and to mention one or two spurious deductive forms which are a source of a good deal of crooked thinking.

1. Hypothetical Arguments

'If *p* implies *q* and *q* is false, then *p* is false.'

This states that if the consequent of an implication is denied, then the antecedent must be false.

There are two things we can do with a hypothetical proposition to obtain a valid inference: (1) We can affirm the antecedent; (2) We can deny the consequent. What we cannot do without falling into fallacy is to deny the antecedent or affirm the consequent. Thus: 'If you are a poet you are a genius' does not imply: 'If you are not a poet you are not a genius'; nor does it imply: 'If you are a genius you are a poet.' But an enormous number of people would seem to think that it does. These two fallacies, *denying the antecedent* and *affirming the consequent*, are exceedingly common.

Denying the antecedent is a fruitful source of error. It is, I suppose, obvious that if *p* implies *q* and *p* is true, then *q* is true; so many seem to think it equally obvious that, if *p* implies *q* and *p* is false, then *q* is false. The mistake is most often seen when an opponent in a discussion is told that his conclusion cannot be true because his premisses are false.

Affirming the consequent is a trap into which all of us fall when we think that, because a consequence of our pet theory is confirmed, the theory must be true. We must suppose that some find it obvious that: 'If *p* implies *q* and *q* is true, then *p* is true.' But they are mistaking it for the valid deductive form: 'If *p* implies *q* and *q* is false, then *p* is false.'

2. Disjunctive Arguments

'If *p* or *q* is true and *q* is false, then *p* is true.'

The 'or' of logic is the weak disjunction which excludes the possibility that both are false though both may be true: 'He is a fool or a knave.' The strong 'or' makes the alternatives exclusive, indicating that the propositions can be neither both false nor both true: 'This is small-pox or chicken-pox.'

The reasoning validated by the formal proposition above is valid even if we do not know which type of disjunction is intended. It is only when we *affirm* one of the propositions and conclude that the other is false that we risk falling into fallacy: for such reasoning is valid only for the exclusive disjunction.

3. The Dilemma

This is an argument from a combination of hypothetical and disjunctive premisses. Of a man with angina pectoris we might say: 'If he remains at work he will remain ill, and if he gives up work he will be a pauper; but he must either remain at work or give it up: therefore he will remain ill or be a pauper.' The form is:—

If *p* then *q*, and if *r* then *s*;
p or *r*: therefore
q or *s*.

It begins with two implications. We are expected to admit that one or other of the antecedents is true: these are the *horns of the dilemma.* And it follows that we must accept that one or other of the consequents is true. Both alternatives are usually unpleasant; and the argument is usually a fallacy.

In another form the minor premiss disjunctively denies the consequent of the implications, the conclusion in the same manner denying the antecedents:—

If *p* then *q*, and if *r* then *s*;
but not-*q* or not-*s*: therefore
not-*p* or not-*r*.

An example is:—

If God is omnipotent then he can prevent evil,
and if he is beneficent then he will prevent evil;
but either he cannot or will not prevent evil: therefore
he is either impotent or malevolent.

It is usually possible to *escape between the horns of a dilemma*, because it is rare for the alternatives to be exhaustive. It is merely necessary to point out that there are other alternatives. The patient could point out that he could do less work, or a different kind of work, and the conclusion would not follow. Another method of escape is to *take the dilemma by the horns*: to show that an asserted consequent does not follow. A theologian might say that it does not follow that if God is beneficent he will prevent evil.

4. Syllogistic Arguments

'If *p* implies *q* and *q* implies *r*, then *p* implies *r*.'

It is not hard to find arguments which, when analysed, can be seen to have this form. Thus, Herbert Spencer, trying to show that egoism may be a source of altruism, argued: 'If men carry self-regard far enough to keep in good health and high spirits implies they become an immense source of happiness to those around [*p* implies *q*], and this implies they become more capable of altruistic actions [*q* implies *r*], then men carry self-regard far enough to keep in good health and high spirits implies they become more capable of altruistic actions [*p* implies *r*].' This in the old logic was called the 'pure hypothetical syllogism'.

It will be seen that the formal proposition validates the categorical syllogism: thus with the minor premiss put first—

	For all *x*'s:—
All Greeks are men:	*x* is a Greek implies *x* is a man;
All men are mortal:	*x* is a man implies *x* is mortal: therefore
All Greeks are mortal:	*x* is a Greek implies *x* is mortal.

5. 'Immediate Inference'

'If *p* implies *q*, then *p* and *r* implies *q* and *r*.'

A proposition *r* can be added to both sides of an implication. 'A horse is an animal, therefore the head of a horse is the head of an animal' is validated by this theorem. Thus:—

(*p*) *x* is a horse	implies (*q*) *x* is an animal,
(*r*) *y* is the head of *x*	(*r*) *y* is the head of *x*,
(*p* and *r*) *y* is the head of a horse	implies (*q* and *r*) *y* is the head of an animal.

In the old logic this was a form of 'immediate inference'. There are numerous examples: 'A Negro is a fellow creature, therefore a suffering Negro is a suffering fellow creature'; or an instance given by Leibnitz: 'Jesus Christ is God, therefore the mother of Jesus Christ is the mother of God.'

Silly fallacies can result from inattention to the exact form: '*Persuasion* is not *Mansfield Park*, therefore the author of *Persuasion* is not the author of *Mansfield Park*.'

A number of deductive fallacies have now been mentioned: deductive reasoning in which the conclusion does not follow from the premisses, although superficially it appears to do so. The form of the inference is not one of the valid deductive forms discussed in the preceding pages, though it closely resembles one. It derives its apparent cogency from this resemblance.

I should make it clear that, in all such logical fallacies, where the conclusion has no logical relation with the premisses, the conclusion is not necessarily false. A faulty argument tells us nothing about the truth or falsity of the conclusion. We cannot therefore disprove a contention by pointing out that the argument is fallacious: we can only say that we have been given no grounds for accepting it.

CHAPTER VII

THE CREDIBILITY OF GENERAL PROPOSITIONS

PURE science is about general propositions.

Each one of us is so constituted that when he repeatedly sees that something has some quality, or that some event has always been accompanied or followed by some other event, he expects these associations in the future. And he makes assertions such as: 'All crows are black'; 'All swans are white'; 'Everyone jumps when touched'; 'Night always follows day'—innumerable assertions of the form 'all α's are β's'.

In the beginning the important thing was to try to find out which of these simple generalizations were true. They were mostly false. As efforts to understand the world increased, more and more men's thoughts turned to the discovery of general propositions of a wider generality, from which simple generalizations could be deduced. We call them 'causal laws'. And as science progressed it became clear that many of these laws could in their turn be deduced from a few laws of much wider generality. For instance, observation made men reasonably confident that anyone who slips off the edge of a cliff falls. Soon it could be deduced from the law that 'all unsupported bodies fall'. The more precise laws of falling bodies were discovered by Galileo. Then it was seen that these laws, with Kepler's three laws of planetary motion, the laws of the tides, and others, are all deducible from Newton's law of gravitation.

A pure scientist's concern is to establish general propositions and to show how they are interrelated in such a deductive system. The aims of applied science, or technology, are the very reverse of this. The applied scientist begins by accepting the laws which others have established; and his interest is centred in particular facts deducible from them, when they help in the solution of his practical problems.

I am concerned at the moment solely with the aim of pure science, which is to understand the world around us.

Our immediate problem is to see what kind of fact corresponds to a general proposition and makes it true; for propositions are said to be true if there is a corresponding fact.

All scientific inquiries start with facts of observation and end with general facts, and the two are entirely different in kind. The premisses with which reasoning starts are singular propositions that can be verified by observation, or could have been so verified in the past. 'This specimen of carbon dioxide is heavier than air' can be known to be true if the fact corresponding to the proposition can be observed; 'Caesar crossed the Rubicon' if we know someone saw him do it. But the conclusions with which reasoning ends are general propositions, and these do not assert the same kind of thing.

'All α's are β's', for all α's, whether observable or not, asserts a necessary association between α's and β's. 'All men are mortal' asserts a necessary association between humanity and mortality; 'All dogs bark' that being a dog entails barking.

A general fact, then, is the association of two things by the relation of implication. In every general proposition we assert this relation. 'All men are mortal' asserts that, for all x's, *if* x *is a man then* x *is mortal.* To know 'all men are mortal' is true, we would have to demonstrate the fact that humanity entails mortality. In general, 'all α's are β's' is true if there is a fact that for all x's, x is an α implies x is a β.

Now this kind of fact is not observable, and yet we have extracted it from facts of observation. What has been the nature of the reasoning process? For we believe innumerable general propositions: life would be impossible if we did not. We believe that if we put a hand in the fire it will be burnt; that if we go out in the rain we shall get wet; that if we slip off the edge of a cliff we shall fall; that we shall die one day. Each is an instance of a general proposition which we unhesitatingly accept.

The reader will have no difficulty in finding premisses from which almost any general proposition can be deduced. 'All men are animals, and all animals are mortal, therefore all men are mortal.' But he will always find that it can be deduced only from premisses which contain another general proposition. The problem is simply pushed back—how do we know the general proposition of these premisses? We may show that it can be deduced from propositions of even wider generality, but eventually we arrive at a general proposition that cannot be deduced from anything. We cannot deduce the law of gravitation from anything.

It is clear enough that we can know general propositions neither by observation nor by deduction.

The kind of reasoning which starts with particular facts and ends with a general proposition as the conclusion is called *induction.* For centuries logicians have tried to validate inductive reasoning, but without success. It seems to be beyond the power of the human intellect to know any general proposition. This means that we can never know anything with certainty.

In deduction, if the premisses are true and the reasoning valid, the conclusion is certain: it admits of no doubt whatever, unless we refuse to accept the primitive axioms of logic. This is not so with inductive reasoning: here, even though we may know the truth of the premisses it does not follow that the conclusion is true; indeed we can never know its truth.

Inductive thinking, however, has one very great advantage over deduction: whereas a valid deductive inference can never go beyond the premisses, giving knowledge not latent in them, inductive inference always goes far beyond the premisses and gives us new knowledge. From observable particular facts we arrive at general principles. It must be admitted that we can never know whether these principles are true, but what we can know is something about their probability.

We cannot be entirely certain that if we slip off the edge of a cliff we shall fall, but we can establish a high degree of probability for it. Our lives are guided by the general propositions we believe. The world seems an ordered place, and we expect that the uniformities we have found in it will continue. We expect that night will follow day; that the tides will go on ebbing and flowing; that water will run downhill and not uphill; that roast beef and Yorkshire pudding will always do us good and not one day poison us; that onion seed will give a crop of onions and not of turnips. We may be wrong. But the thought has little influence on our daily lives. As Bishop Butler said: 'Probability is the very guide of life.'

We shall not get very far with this unless we are clear about the two concepts, both called 'probability'. It is now universally recognized that 'probability' names two entirely distinct notions. One refers, in Carnap's phrase,* to 'degrees of confirmation' of a general proposition; the other to mathematical probability, or 'relative frequency'. Each is susceptible of a number of interpretations, but it was the inability to separate the two fundamental ideas that brought the bewildering confusion which vitiated all discussion of the meaning of 'probability' until quite recent years.

The two following propositions are quite different in meaning:—

1. It is probable that any α is a β.
2. It is probable that all α's are β's.

'It is probable that any α is a β' is a statement about mathematical probability: it says something about the chances of an α being a β when the two are fortuitously associated. It may be the case that, while an α may or may not be a β, α_1 is a β. There being no necessary association, we say it is a matter of chance that α_1 is a β. A card may or may not be an ace: we say it is a matter of chance that a card drawn at random from an ordinary pack is an ace. Of course, what card is drawn is determined by something: it is determined by the combined action of numerous unknown causes, and it is this that we call 'chance'.†

The *probability of an event* is a fraction representing the relative frequency with which the event occurs in the long run. Thus in every 52 times a card is drawn at random from an ordinary pack, an ace will be drawn 4 times in the long run. The probability of a card drawn at random being an ace is therefore expressed by the fraction $\frac{4}{52}$. By 'long run' we mean 'very long run': that if we never stop drawing cards at random, the frequency with which an ace is drawn will approximate more and more closely to 4 in 52.

In general terms, if α is a class of n members, m of which are also β's, the probability that an α taken at random is a β is m/n.

A statement about the mathematical probability of an event is not something that is 'probably true': it has all the certainty of a mathematical theorem. Mathematical probability is not concerned with human doubt or uncertainty.

'It is probable that all α's are β's' is a statement about 'probability' used in a different sense: it assigns *credibility* to a general proposition asserting necessary association of α's and β's.

---

* Carnap, R. (1951), *Logical Foundations of Probability*. London: Routledge & Kegan Paul.

† The meaning of chance and mathematical probability are considered in more detail in Chapter XVII.

The most lucid analysis was by Maynard Keynes; and his interpretation is now the most widely accepted.*

A proposition can only be true or false. When we say, not 'It is true that——', but 'It is probable that all α's are β's', we are saying something about the evidence that bears on our knowledge of its truth or falsity. There is an ellipsis; the correct statement is: 'There is a certain probability upon the evidence that all α's are β's.' We are saying that the evidence we have gives the proposition a certain degree of credibility.

It is not contradictory to say that a false proposition is probable upon the evidence, or that a true proposition is most improbable. I think it was the late Mgr. Ronald Knox who used to say that he liked his religion to be true even though improbable. Clearly the degree of probability, or credibility, given to a proposition by the evidence changes as the evidence changes.

So far the notion is fairly straightforward. We all recognize that certain evidence will give a rational man a degree of confidence in the truth of a proposition, on which he is wise to act. He cannot demonstrate with the rigour of deductive logic that he will fall if he steps off the edge of a cliff: but the evidence is such that he is wise to act on the credibility it gives to the proposition.

Things become more difficult when we ask: What is the meaning of 'degree of credibility'? The degree of credibility of a general proposition cannot be measured as can mathematical probability. We can say that 'All dogs bark' has a high degree of credibility, but the notion is totally different from that of a degree of mathematical probability. We are not thinking of the probability that any dog taken at random barks: we are thinking of the degree of doubt with which our belief 'all dogs bark' is tinged. And how do you measure a degree of doubt?

This depends on the kind and amount of evidence on which the belief is based. We can say of some general proposition that the evidence is such that its credibility is zero—meaning it is certainly false; or we can say that the evidence is such as gives it a credibility which approaches unity—meaning it is almost certainly true. But we cannot weigh intermediate evidence in such a way that we can assign to it a fraction intermediate between 0 and 1, as we can with mathematical probability. And if we could, confusion between the two 'probabilities' would be even more confounded. The best we can do is to talk about a low, or a high, or a moderate degree of credibility.

In the following chapters we shall consider the ways in which a general proposition can be shown to have credibility, and how the credibility assignable to it may in some instances be increased until it approximates to certainty.

---

* Keynes, J. M. (1921), *A Treatise on Probability*. London: Macmillan.

CHAPTER VIII

INDUCTION BY SIMPLE ENUMERATION

'ALL swans are white' was believed by all Europeans for many centuries, because it seemed that every swan anyone had ever seen had been white. This kind of reasoning is called *induction by simple enumeration.* At first sight it appears to be: All α's we have observed have been β's, therefore all α's are β's. But such reasoning is clearly absurd: there is no logical principle that makes it valid. If I have known, say, six red-haired people and they were all bad-tempered, I cannot conclude: 'All red-haired people are bad-tempered.' The conclusion would be falsified by the first contradictory instance, as 'all swans are white' was proved false when black swans were found in Australia.

And yet we have reasoned in this way since the dawn of human intelligence. Has it no cogency at all? At least, we would say, you must admit that it is *probable* that all red-haired people are bad-tempered. Moreover, we are convinced that the greater the number of times the association has been observed, the more certain we can be of the conclusion.

What we have to consider is whether there is an inductive principle that validates probability inferences. Bertrand Russell* says that if there is such a principle, it might be expressed: 'Given a number n of α's which have been found to be β's, and no α which has been found to be not a β, then the two statements: (1) The next α will be a β, (2) All α's are β's, both have a probability which increases as n increases, and approaches certainty as a limit as n approaches infinity.'

It is not, however, immediately clear how we get this idea of the influence of numbers of instances. If I join two terminals from a battery, with a galvanometer in the circuit, to the ends of a length of pure copper wire, I can show that the copper conducts electricity. But I cannot infer from that simple observation: 'All copper conducts electricity', although this general proposition does acquire some small degree of credibility. Is this increased if I repeat the experiment and get the same result? If I repeat it with the same piece of copper wire or with a closely similar piece, under precisely the same conditions, nothing is gained. If I repeat it a thousand or ten thousand times, I add nothing to the credibility given by a single observation.

Perhaps a clue may be found in the undeniable fact that we feel far more certain of the truth of some general propositions than of others. We do not doubt for a moment that if we slip off the edge of a cliff we shall fall, for we accept without question the truth of the general proposition: 'All unsupported bodies fall.' We do not feel so certain of 'all dogs bark', though a burglar presumes that they do. We may think it probable that all red-heads are bad-tempered, but we should not be astonished to find one who was not.

---

* RUSSELL, B. (1948), *Human Knowledge*, Chapter VII. London: Allen and Unwin.

Whenever it seems absurd to question the truth of some general proposition it will be found that we have a vague impression of *necessary* association. It seems certain, from innumerable instances of the association, that all α's have been found to be β's for some reason, and it seems probable that the cause lies in the essential nature of the thing that we classify as an α.

We might tentatively accept 'all crows are black' because we think that there may be some cause for it in the essential nature of a crow. We would accept 'all copper conducts electricity', not just because all samples so far tried have done so, but because it seems probable that there is something in the essential nature of copper that makes conduction of electricity a necessary property of it. We would accept 'every animal that has lungs has two ventricles to its heart', not because of observed instances, but because we know why. We believe that if we slip off the edge of a cliff we shall fall, not because everyone who has done so has fallen, but because we believe that there is something in the essential nature of matter that explains why all unsupported bodies fall.

What simple enumeration does is to suggest a causal connexion of this sort, and to lend a varying degree of support to it. This support is derived from the variety and not from the number of observed instances.

To return to the repeated demonstration that a given piece of copper conducts electricity—a critic of the conclusion 'all copper conducts electricity' might say: 'How do you know that with copper wire of some different calibre, or different length, or at some different temperature, conduction might not cease?'

I can answer this only by repeating the experiment, not with the same or a similar piece of wire, but with wire of different lengths and calibres, at different temperatures, and so on. I cannot test copper in all possible shapes and at all possible temperatures, but the oftener I repeat the experiment under different conditions and get the same result the more probable it becomes that variables are without effect, and that conduction of electricity is associated with what remains constant in all the observations, namely, copper.

Maynard Keynes was the first to show that it is not increase in the number, but increase in the variety of instances, that strengthens the credibility of an induction.

Let us say that I am trying to show that all men are mortal. I collect information about 100 men born, say, over 110 years ago, and find that they are all dead. A critic might say: 'But all your men were from the London poor; perhaps they died not because they were men, but because of the fortuitous circumstance of poverty.' I counter this by studying further instances, taken now from all sections of the community and from all parts of the country, and I still find that all men born over 110 years ago are dead. The critic then suggests that this is because they were all Englishmen, and not because they were men; perhaps some men in Tibet are immortal. I therefore increase the instances to include all groups of Europeans, Asians, and Africans, and still I find the same association of humanity and mortality. The credibility of 'all men are mortal' is progressively strengthened by increasing the variety of instances, because the effect is to eliminate characters fortuitously associated with those essential to being a man, and that might account for the observed mortality.

Again, if we see that every sort of thing, and not merely a large number of things, falls when unsupported, it becomes clear that falling is associated with just the fact that they are all material bodies.

Usually, to increase the number of observations is to increase variety. And this seems to explain the conviction we have that the credibility of a generalization increases with the number of observations on which it rests. But to get enough variety to give the conclusion more than a moderate degree of credibility the number of observations would have to be enormous.

The credibility of an inductive conclusion can, of course, be sharply increased if we are prepared to substitute 'most' for 'all'. And in fact it is really propositions like 'most dogs bark' or 'most people jump when touched' that guide our daily lives. But scientists are not content with this kind of knowledge, except as stepping-stones in the discovery of laws to which there are no exceptions whatsoever.

CHAPTER IX

THE INSPIRED GUESS

It will come as a surprise to many people that what appears to be pure guesswork is an essential part of the method of inquiry called 'scientific'.

The principle of induction by simple enumeration, which from now on I propose to call simply 'induction', confers some degree of credibility on the generalizations with which science begins. The instances that exemplify these 'laws' are the facts which have been directly observed: it was these straight observations that suggested the generalization. Any barking dog is an instance of, and suggested, the generalization 'all dogs bark'.

But when a particular science advances, many of the laws that are established have instances that are observable only under experimental conditions. It is only when circumstances are deliberately arranged that the facts which exemplify the law become observable. It is obvious, therefore, that the law was thought of first and the facts observed later. Unless it were to test the truth of the law it is extremely unlikely that anyone would have made the observations. Some mental process other than induction must therefore be at work in establishing laws of this kind.

Laws with observable instances—observable without the aid of experimental techniques—are established by induction. Examples abound in the early history of any science.

Karl von Baer (1792–1876) came to the study of embryology when the generation of living things was almost as great a mystery as it had been to Aristotle. With the enormous zest and ambition of youth he planned to reveal all the details of the embryo's development in all vertebrates, if not in all animals. In his student days he had thought nothing of walking 400 miles for the privilege of studying for a few months under an admired teacher.

He made known a mass of new facts, from which several inductive generalizations of the highest importance emerged. 'All through his slow toilsome study of the folds and twists, the thickenings and thinnings of the growing chick', T. H. Huxley wrote in his obituary notice, 'he was supported by the sure hope that in the flitting shadows of embryonic forms was to be found the key to the laws of animal organization.'

In 1828 Baer wrote:—

> While I was looking at the ovary of a bitch solely with the aim of comparing the structure of this organ with the ovaries of other animals examined by me . . . I saw wholly without care and with unaided eyes a little yellow body in each Graafian vesicle in the ovary, which declared itself as an ovulum under the microscope. The presence of this egg within a Graafian vesicle is probably raised above all doubt for all time. I have found it in all mammals investigated, in the smallest as well as in the largest. . . . After the envelope which surrounds the ovary as a sac has been cut open, a good eye will probably recognise a yellowish dot in most Graafian vesicles. It is the egg.*

---

* 'Commentary on the Communication: On the Eggs of Mammals and Man', translation in Mayer, A. W. (1956), *Human Generations*. California: Stanford University Press.

He found an egg first in dogs; then he looked for it in other mammals—pigs, cows, sheep, rabbits, stags, porpoises, dolphins, man. And so came the inductive generalization that all mammals come from eggs.

Baer's germ-layer theory, a generalization that has so illumined our understanding of the generation of living things, is another inductive inference from observation. It was already known that the tip of a growing shoot of a plant is composed of uniform structureless tissue from which the characteristic structures of leaves and flower parts gradually emerge. Baer showed that the same sort of thing occurs in the development of the chick. After 24 hours' incubation of the egg, three distinct layers of undifferentiated tissue are recognizable. These layers, since called ectoderm, mesoderm, and endoderm, are distinguishable from one another, but each is formless in that no sign of any organ can be discerned in it. The various organs of the body are developed from them, and in all animals that produce these layers each layer gives rise to the same organs: the ectoderm, the skin and nervous system; the mesoderm, the muscles, skeleton, and kidneys; the endoderm, the digestive organs. In all vertebrate animals—whether fish, frog, lizard, snake, bird, cow, sheep, dog, rabbit, porpoise, man, or any other mammal—this is how the embryo begins its growth from the fertilized egg. This important generalization is simply an inductive conclusion from a long series of careful studies of what in fact can be seen to take place.

In his *Descent of Man*, Darwin has an illustration showing a drawing of a human embryo of about 3 weeks' growth, side by side with the embryo of a dog at the same stage. It is hard to tell which is which. In the neck of both are gill-like slits like those of a fish embryo. The heart in both is a single pulsating vessel. In the human embryo the coccyx projects beyond the rudimentary legs like the true tail of the dog. It was an instance of another great generalization that had come from Karl von Baer's unremitting observations, and one which had provided Darwin with more evidence in favour of his theory of descent. We call it now Baer's *biogenetic law*. Comparing vertebrate embryos he had found, without any exception, that the embryos of higher and lower animals resemble each other more and more closely the younger they are; the embryos of related animals are very closely similar in their early stages. It was a simple induction that has proved of fundamental value in biology.

•

Now let us think of how, in the early part of our century, physiologists established the principle of the chemical mediation of nerve impulses. The procedure is reversed. They thought of the law first, then looked for the facts. About 1904 a very curious fact was noticed: that the effect of artificially stimulating sympathetic nerves, on the heart, the blood-vessels, and various glands, was exactly mimicked by administration of an extract of the adrenal gland, called *adrenine*. It was suggested, and now that we know it may seem a fairly easy guess, that sympathetic nerves produce their effect, whether quickening of the heart-beat or whatever it may be, by first producing adrenine at or near their endings.

There may be some reader who has not heard of Loewi's experiment, and who thinks it should not be so difficult to detect this, if it is actually what happens. If so, let him say how it should be done.

It was an obvious extension of the idea to suggest that there is such an intermediate chemical step between *all* nerve impulses and the activity they produce in their effector organ. But no one could have observed it by chance. The chemical mediator produces its effect and is destroyed, in some instances within a fraction of a second.

The law was thought of first: a guess which could explain a striking observation. Then the facts by which it is exemplified were observed in a series of ingenious and difficult experiments, perfected over about a quarter of a century.

If the reader is not convinced that the reasoning is different from that of a straightforward induction, let us pass on to a third kind of law: one about abstract notions which are not observable. Such laws have no observable instances, and therefore could not be arrived at by induction. Atoms and subatomic particles, and quanta of radiation, are not observable. The special laws of quantum theory for single atoms and their quanta of radiation were not therefore revealed by inductive reasoning. The law of gravitation is usually given as an example of this kind of law. It states: 'Every particle of matter in the universe attracts every other particle with a *force* whose direction is that of the line joining the two, and whose magnitude is directly as the product of the masses, and inversely as the square of their distance from each other.' It is a law about 'forces'. 'Forces' are not observable. The law of gravitation was therefore revealed by a method of reasoning other than that of induction.

The technical name for a guess at the explanation of some unaccountable fact is *hypothesis*. It means simply 'supposition'. It is in fact the same word, from ὑπό=under, and θέσις=placing; the Latin *suppositio* meaning 'a placing under'. A hypothesis is a suggested underlying causal law of which the fact to be explained is a consequence. A strange fact is explained or made clear when it is shown to be a consequence of some law.

It is said that no one can tell you how to think of a hypothesis. But it may be possible to give some useful hints.

The mental transition from a particular fact to an imagined law which would explain it seems like a leap in the dark. It is a leap most people are ready to risk, whatever it is that needs explaining. But their guess is usually wildly improbable. If we want to get some idea of how fruitful hypotheses are made, we might do worse than consider first the nature of the spurious, the useless, and the improbable kinds.

Not every explanatory guess is a scientific hypothesis. There is one indispensable requisite: it must have verifiable consequences. This means that we must be able to deduce from it that certain facts will be found if we look for them. If we suggest that a patient subject to 'attacks' suffers from epilepsy, that is a hypothesis, for we can deduce certain things from it and see if they accord with further observation. But a suggestion that he may be possessed by an evil spirit is not a scientific hypothesis: you can deduce nothing from it that could be compared with the results of observation and found to agree or disagree with them.

All teleological explanations fall into this category of spurious hypotheses. It has nothing to do with religious belief. Even if it were shown in some other way that cancer is a punishment for the sins of mankind it would still not be a scientific hypothesis: yielding no verifiable deductions it could not be examined by the scientific method of inquiry.

Hypotheses are almost as useless as the guesses that have no consequences, if we can deduce from them only facts that would be impossible to observe with known experimental techniques. But perhaps Michael Foster's judgement was a little too harsh. In his life of Claude Bernard he delivers himself of this mordant outburst: 'The man who constructs a hypothesis without supplying an adequate programme for its trial by experiment is a burden to science and to the world; and he who puts forward hypotheses which by their very nature cannot be so tried, is worse, for he is a purveyor of rubbish.'*

Laymen are very fond of telling the medical profession what they think is the probable cause of an obscure disease, or what method of treatment should remove an intractable complaint; and they can be very annoyed when little or no notice is taken of them.

Most popular explanations of things obscure to the experts are seen almost at once to be highly improbable, usually because they can be seen to be in conflict with some well-verified theory. If such a hypothesis were true the theory would be false, and the evidence for this may be such as to give it a very high degree of credibility. It follows that the suggested hypothesis is improbable. One, for example, involving the inheritance of acquired characters, would be improbable, because if it were true the theory of the gene would have to be abandoned or much modified.

But to be thought improbable a hypothesis need not clash with one of the fundamental laws of the science; it is enough if it conflicts with knowledge which an expert believes to be already well established. A supposition that cancer of the stomach is caused by the use of aluminium saucepans could be investigated statistically, with expenditure of considerable time and patience and money. But if an expert believes that it conflicts with what is already known of the characters of carcinogenic substances, why should it be? It is inherently improbable.

Again, as will be shown in Chapter X, those hypotheses that begin life with the highest degree of probability are those that can be seen to explain far more than the immediate obscurity. These are the fruitful hypotheses, revealing whole tracts of new knowledge, predicting facts hitherto unknown and wholly unexpected. The layman forgets that the expert may have half-a-dozen or more hypotheses of his own, each more probable than his. Of a sort they are not difficult to make. As Pasteur remarked: 'Hypotheses come into our laboratory in armfuls.'

If it can be seriously suggested, as once it was, that the cause of stomach cancer is the use of aluminium saucepans, why not anything? Why not suggest eating boiled cabbage, smoking pipes, listening to classical music—anything—and expect the Medical Research Council to follow it up? We would all say that there is no rhyme or reason in such suggestions. And we would be right.

It follows that they must be wrong who say that scientists are not guided by any principles of reasoning, but only by their imagination.

* FOSTER, M. (1899), *Claude Bernard.* London: T. Fisher Unwin.

There must be logical stepping-stones to a hypothesis, even if we cannot see what they are.

Most hypotheses are reached by one of the weakest of all arguments: *argument from analogy*. Two phenomena, A and B, resemble one another closely in some respects: x is known to explain A, therefore something like x probably explains B.

Darwin has told us how his theory of descent with modifications, through variation and natural selection, came into his mind. The central hypothesis is the *law of natural selection*, the most important generalization in biology. It occurred to him as the probable explanation of the problem: 'How is it that varieties become ultimately converted into distinct species?' His supposition was that those individuals that survive the struggle for existence and procreate are varieties with some advantage over their fellows, whereas varieties less fitted for survival are gradually eliminated. The same struggle for existence awaits descendants of the surviving varieties, and so there will, over the generations, be a slow accumulation of beneficial modifications. Over a long period a new species can be expected to emerge by the preservation of small improvements generation after generation. Over many millions of years the descendants of a creature of the reptile type might be so altered that we would class them as mammals or birds. 'This preservation of favourable individual differences and variations, and the destruction of those that are injurious, I have called *natural selection*', he wrote. 'How extremely stupid not to have thought of that', Huxley remarked.

But Darwin thought of it because he had been wrestling with the problem for some 15 months, when he happened to read a well-known essay, then in its sixth edition, by Robert Malthus, a former professor of political economy. Its theme was that population tends to increase faster than the means of subsistence. Unless we restrict our rate of increase, he warned, war, or disease, or hunger will check it; and this will always be an obstacle to a future age of equality.

Because his mind was already on the problem of the evolution of species Darwin at once saw the analogy. He was well aware of the universal struggle for existence among living things; if this would prevent human equality, so it would throughout the animal kingdom: 'It at once struck me that under these circumstances, favourable variations would tend to be preserved, and unfavourable ones to be destroyed. The result of this would be the formation of a new species. Here then, I had at last got a theory by which to work.'

It may well be that all original discoveries come from seeing analogies not seen by others. But a number of illustrious pioneers in science have insisted that there are no logical steps to a hypothesis. When it comes, it comes spontaneously into consciousness. Usually it has come inexplicably after prolonged and intense thought. They have ascribed it to intuition or imagination. Poincaré, the mathematician, physicist, and philosopher, said that logic had very little to do with discovery. Einstein said of physical and mathematical laws that there is no logical way to their discovery—there is only the way of intuition. Max Planck, the originator of the quantum theory, said that what is needed in attempting to find order in apparent chaos is imaginative vision, not reason.

Until very recent years it was a popular belief that scientists are a different race of men from scholars in the arts and humanities. Their way of thought was held to be different in kind. Now the belief is growing that this is an error. So far from being intellectual strangers they share one guiding light. Accompanying, if not preceding, Reason and Revelation a third angel, Imagination, guides the path of man, scientist and artist alike, in his endless search for truth.

The faculty called 'imagination' is the ability to reproduce mental copies of sensations once experienced. According to William James, 'The imagination is to be called "reproductive" when the copies are literal; "productive" when elements from different orginals are recombined so as to make new wholes.' If you can hold in your mind's eye a clear image of a crystal, or of any of the diagrams in a textbook of solid geometry, you have reproductive imagination. The productive kind is what is usually meant: an imagery of possible sensations, reproducing no past combination exactly.

Kekulé's dream is an instance of productive imagination. Kekulé was the German chemist who, last century, first suggested the benzene-ring formula which has been so fruitful in organic chemistry. He described how this hypothesis came to him. He was sitting by his study fire, half asleep, and saw long chains of carbon atoms wriggling like snakes. Suddenly one of these snakes seized its own tail. He awoke with a start, and there was the benzene hexagon.

It seems unlikely that pictorial thinking is how most scientific hypotheses have been generated.

William James, in *The Principles of Psychology*, records at some length an inquiry into how individuals differ in visual imagination, carried out by that many-sided genius, Sir Francis Galton, in 1880.

Galton began a series of questionnaires by asking some of his scientific friends to tell him details of their mental picture of the morning's breakfast table: 'To my astonishment I found that *the great majority of the men of science to whom I first applied protested that mental imagery was unknown to them*, and they looked on me as fanciful and fantastic in supposing that the words "mental imagery" really expressed what I believed everybody supposed them to mean. . . . They had a mental deficiency of which they were unaware, and naturally enough supposed that those who affirmed they possessed it were romancing.' One of these scientific friends wrote to him: 'It is only by a figure of speech that I can describe my recollection of a scene as a "mental image" which I can "see" with my "mind's eye". . . I do not see it.' On the other hand, many men whom he met in general society, a large number of women, and many boys and girls assured him that they habitually saw mental imagery, and that it was perfectly distinct to them and full of colour.

'My own conclusion', Galton wrote, 'is that an over-ready perception of sharp mental pictures is antagonistic to the acquirement of habits of highly generalized and abstract thought, especially when the steps of reasoning are carried on by words as symbols. And that if the faculty of seeing the pictures was ever possessed by men who think hard, it is very apt to be lost by disuse.' But he says later: 'Men who declare themselves entirely deficient in the power of seeing mental pictures can nevertheless give life-like descriptions of what they have seen, and can otherwise express themselves as if they were gifted with a vivid visual imagination.'

Many people do think in pictures. We all do in dreams. And pictorial thinking can undoubtedly help in the solution of problems. Arthur Koestler, in his popular work *The Act of Creation*, tells a little story which illustrates this. He describes how he put the following problem, taken from the *Scientific American* of June, 1961, to a number of his friends:—

At sunrise a Buddhist monk begins to climb the spiral path leading to the temple on the summit of a tall mountain. He climbs now quickly, now slowly, stopping many times on the way, arriving at the temple shortly before sunset. He stays in the temple for meditation. Some days later he descends along the same narrow path, again starting at sunrise. Again his pace is now fast, now leisurely, and he stops frequently. Even so, his average speed is greater than when climbing. Prove that there is a spot on the path where the monk will be on both trips at precisely the same time of day.

A few trained scientists tried an approach from well-known mathematical principles, and failed to find a proof. Some said it was not true. But a young lady, aided only by her imagination, solved it immediately. For she *saw* the monk—he wore a saffron robe—wearily plodding upwards. And she *saw* him coming down, not a week, or whatever time it was, later, but at the same time. It did not worry her at all that the same man cannot be both climbing and descending at the same time: he was—she saw him. And she saw that he must meet himself at some time, and at some spot on the path, which, of course, is the spot required.

There is, however, no reason to distrust Galton's conclusion that most scientists are mentally deficient in the faculty of visual imagination. And it is hard to believe that all his scientific friends were of the kind who would never make any original discovery. The probability is that most hypotheses consciously arrived at come through the ordinary processes of associative thought, where visions need not be visual. Perhaps, after all, this is what most people who talk of 'imagination' mean by it.

Those who say that logic had very little to do with their discoveries must mean that they have not the faintest idea how they came to think of them. They cannot mean that straight thinking is no more likely to produce results than is the crooked stuff that dreams are made of.

Some of the most illuminating and fruitful hypotheses have seemed like inspired guesses. They have leaped into consciousness spontaneously and unexpectedly. They are the conclusions of unconscious thought.

Nothing compels us to believe that there can be no purposive thinking unless we are aware of it. Probably most of our cerebration goes on unconsciously, a result now and then coming up into consciousness as an apparently spontaneous thought, or being pulled up by a conscious mental effort. And nothing compels us to believe that all unconscious thought is the primitive pictorial thinking of dreams.

The unconscious mind will continue to work at a problem when consciously we have stopped thinking about it. So many of those whose thought has illumined the world have confirmed the experiences related by Henri Poincaré in his *Science and Method*.

There is a period of concentrated but fruitless work on a difficult problem. For one reason or another it is stopped and attention is given to other things. Then, after a longer or shorter time, and when not prepared for it, a sudden illumination comes unasked, accompanied by a strong feeling of conviction. Or sometimes the answer has sprung into consciousness very soon after resuming work on the problem.

Describing the progress of a series of mathematical researches, Poincaré tells how, during a night of insomnia after several days of fruitless effort, the solution came suddenly. Another time a key hypothesis came to him during a conference, when his thoughts were elsewhere. He had many similar experiences. Having had no success with another problem he left it and went to the seaside: the solution came suddenly, and with the usual feeling of conviction, while walking along the cliff.

It is impossible to escape the conclusion that the period of rest has been occupied by unconscious work on the problem. Moreover, it seems clear from recorded experience that this unconscious thought is not possible unless it is started by a period of conscious work. Sudden inspirations never come unless there has first been a spell of conscious effort.

It may be that unconscious thought is capable of results that conscious thought cannot achieve. Poincaré believed that the unconscious mind could pass in review an incredible number of possibilities, and allow to pass into consciousness the one both useful and, in a sense which mathematicians and scientists will understand, elegantly beautiful. We may picture the unconscious mind considering a succession of hypotheses, judging and rejecting, And we may assume that the inspired guess is the one allowed into consciousness because it excites admiration by its elegant simplicity and promise. If two or three hypotheses occur to us, they are the two or three that unconscious thought has 'short-listed'.

This is a gift that many people do not know they have, though it is probably common. It is common knowledge that we can with advantage 'sleep on a problem': that the solution may come clearly and suddenly into the mind next morning. But few seem to know that unconscious thought on a very difficult problem may continue for days or weeks, once it has been set going by a period of conscious thought, and many scientists have told how much they rely on it.

Can this faculty of unconscious thought be cultivated? The only help that can be given is what we can gather from the recorded experiences of men of the calibre of Henri Poincaré, W. B. Cannon, Von Helmholz, Metchnikoff, Gauss, Loewi, and many others. It is clear that inspiration will not come 'out of the blue'. The unconscious mind will not think about a problem unless the mechanism is started by a spell of concentrated voluntary work. If no possible hypotheses come into the mind during this conscious work, leave the problem alone for a time: turn the mind to other interests, or go away for a holiday. Then if still no solution has come, resume conscious work.

The physicist, Sir George Thomson, writes of his even more celebrated father, J. J. Thomson: 'I have often heard him say that the best way to conquer a difficult problem is first to study it hard, and then go away and think of something else; when one got back the solution often fell out.'

CHAPTER X

SCIENTIFIC METHOD

'I HAD at last got a theory by which to work.' Darwin was using 'theory', as very many people do, for 'hypothesis'; the word has other meanings, but this is its popular usage. Working by a theory is the scientific method of inquiry.

Everyone who invents a theory is expected either to undertake the work of verification himself, or to keep quiet about it. Otherwise he is what Michael Foster said—a burden to science and the world. If he is engaged in research, convinced though he is of its truth, he has to find evidence that will convince others. This, of course, is research: most of the facts of any science have been discovered in the course of verifying some theory.

If Reason ever did hand over the torch to Imagination, it is taken back now. The research worker looks for the evidence he wants: Reason will determine whether he has found it. Even if he thinks he believes that logic has few discoveries to its credit, he will plan his work of verification in the light of the angel's torch.

And the angel's principle is that what cannot be proved untrue may be accepted as true.

As an example of scientific method in action we may take the verification of the inspired guess that has proved to be the foundation of modern genetics.

In the gardens of Altbrünn monastery, in what is now Czecho-Slovakia (Brno), in the year 1856, a strip of ground was allotted by the prelate to a young priest who wished to carry out some botanical research. This amateur was Johann Mendel, a peasant's son, who had taken the monastic name 'Gregor'.

A strange idea had occurred to him, that could explain a remarkable phenomenon often observed in the hybridization of plants. If it were true, it might be a key to the seemingly insoluble problem of how characters are transmitted from parent to offspring.

In the 1850's it was known that sexual reproduction begins by the fusion of two germ cells—technically called 'gametes'—one from each parent.

It was known that this was true of plants as well as animals. In flowering plants what happens had been closely watched and elucidated in 1846 by Giovanni Battista Amici, a brilliant astronomer and mathematician among his other attainments. He had shown that an egg is present in the ovule of a flower, and that it is stimulated to develop into the embryo, or seed, by the male gamete that passes along the pollen tube to it.

It was therefore beyond dispute that every feature of the complex anatomical structure of the parents, with all their thousands of peculiarities, their aptitudes, idiosyncrasies, traits of character, even such trivial things as colour of hair and eyes and skin, and tricks of speech, must in some unimaginable way be carried to the next generation by a single microscopic cell.

The phenomenon that exercised Gregor Mendel's imagination was an observation repeatedly made that plant hybrids, as a rule, are not exactly intermediate between the parents. In many cases he himself had found that one of two contrasting parental characters is so preponderant in the hybrid that it is difficult, or quite impossible, to detect the other.

Paternal and maternal characters are, in fact, in both animals and plants, often inherited as a blend: as when black and white races intermarry and the children are more or less coffee-coloured. This is what might be expected: what is remarkable is that, when there are two such contrasting characters in the parents, offspring may take wholly after one parent or wholly after the other. The child of a brown-eyed father and a blue-eyed mother has either brown or blue eyes, and never a blend of the two.

This phenomenon is very striking when varieties of the garden pea are crossed.

Everyone knows the variety that grows to 6 or 7 feet in height, and the contrasting dwarf variety that scarcely needs staking. The dwarf plants always breed true, and there are true-breeding tall plants; i.e., a dwarf plant self-pollinated—the natural method in peas—produces only dwarf plants; a true-breeding tall plant, only tall plants.

When a true-breeding tall plant is artificially crossed with a dwarf plant, pollen from one being taken to the flower of the other, the hybrids produced are not intermediate in height, nor are some tall and some dwarf: they are all tall.

Mendel's choice of *Pisum sativum* for testing his theory was both very shrewd and extraordinarily lucky. It is a plant very easy to grow out of doors. There are a number of varieties which differ in very easily recognizable contrasting characters, whose fate in the hybrids he could follow. There are, for instance, varieties producing only round 'peas', others only wrinkled; there are varieties producing only green 'peas', others only yellow; and so on. It is an easy plant to pollinate artificially. There is no diminution in fertility in successive generations of hybrids, as there very often is when distinct species of plant are successfully crossed. For all these reasons it was a shrewd choice. It was lucky because many other plants would have given him equivocal results, and he would almost certainly have given up the whole research.

Before beginning the series of experiments to verify his theory, he devoted some two years to securing the true-breeding varieties he would need, to perfecting his technique of artificial cross-pollination, and getting a general idea of the sort of thing that happened. Although entirely without training in research, Mendel never put a foot wrong.

He would take, for instance, a true-breeding variety forming *round* seeds and cross it with a variety forming *wrinkled* seeds: the seed pods that followed contained only round seeds—no single wrinkled seed was to be found. In the following year he planted out these hybrid round seeds and allowed the plants that grew to self-pollinate in the ordinary way. In the resulting seed pods the seeds of the second generation of hybrids were some round and some wrinkled. The character of a wrinkled seed, lost in the first generation of hybrids, reappeared in the second.

Characters transmitted to the first hybrid generation Mendel called *dominant*; those which became latent, to reappear in the progeny of the hybrids, he called *recessive*.

His theory was this:—

It is a safe assumption that a character transmitted from one generation to the next, such as roundness of seed, must in some way be carried in the gametes (egg cells and pollen cells). Mendel had not the faintest idea of what was carried—obviously not the character itself, but it had to be something. It was something responsible for the development of the character. At that time it was not, of course, an observable entity, and he was content to call it by a name later translated 'a factor'. There was a factor for roundness of seed, one for tallness, and so on. Moreover, these factors must be present in both egg cells and pollen cells. This was proved by an early discovery of his that which one of the two plants crossed was selected as the seed-bearing plant made no difference to the result.

It follows that, in the fertilized egg, these factors occur in pairs, one from each gamete. The fertilized egg of a true-breeding round-seeded plant must contain two factors for round seeds; that of a true-breeding wrinkled-seeded plant two factors for wrinkled seeds. And so every cell of the plant body that grows from the fertilized egg, *including the cells from which the gametes are formed*, must contain the same pair.

Mendel's principle, since called the *principle of segregation*, or the *Mendelian principle*, asserts that, in the formation of the gametes, the pairs separate: one factor goes into one pollen cell, the other into a second; or one into one egg cell, the other into a second.

The gametes of a pure-breeding round-seeded plant contain one factor for round seeds; those of a wrinkled-seeded plant one factor for wrinkled seeds; and so, when the two are crossed, the fertilized egg contains one factor for round seeds and one for wrinkled seeds.

Let us call the factor for a dominant character, such as roundness of seed, 'A'; that for the corresponding recessive character, the wrinkled character, 'a'. We can represent what happens when a plant with round seeds is crossed with a plant with wrinkled seeds in this way:—

	Round Seeds	Wrinkled Seeds
Parents	A A	a a
Gametes	(A)	(a)*
F_1	A a	

Set out like this it shows that the pair of factors in the first hybrid generation, the F_1 generation, must be 'Aa'. 'A' being dominant they must all be round seeds.

* It is a common device to show gametes by a circle around the factor concerned, the male with a tail.

Now see what happens to this pair of factors when the F_1 hybrids make their egg cells and pollen cells. Mendel's principle asserts that they separate, one going into one egg cell, the other into another; the two kinds of eggs, 'A' and 'a', will be formed in equal numbers. Similarly, in the formation of the pollen cells: the pairs separate, and the pollen cells receive one or the other, again the two kinds in equal numbers.

When, therefore, an F_1 hybrid self-pollinates, the fertilized eggs will not be all the same. The F_2 generation from the self-pollination of the F_1 hybrid arises in this way:—

	Pollen Parents	Seed-bearing Parents
Parents	A a	A a
Gametes	(A) (a)	(A) (a)
F_2 in equal numbers	AA Aa aA aa	

If the different kinds of gametes, both male and female, are in equal numbers, and any egg cell has an equal chance of being fertilized by any pollen cell, the possible combinations in the fertilized eggs will occur on the average in equal numbers. What these possible combinations must be can be seen most readily by constructing a square like this:—

		Female gametes A	Female gametes a
Male gametes	A	AA	Aa
	a	aA	aa

Three of the combinations, it will be seen, are dominant, and one is recessive. Wrinkled seeds will appear again in the F_2 generation. Not only this; the theory tells us in what proportion. There will be 3 round seeds to 1 wrinkled.

This is what Mendel always found. In one experiment, from 253 F_1 hybrid round seeds which were planted out, the plants that grew yielded 7324 seeds, of which 5474 were round and 1850 were wrinkled—a ratio of 2·96 : 1. Note how he saw the need for large numbers. Each pod, as a rule, contains both kinds of seed, but it would be foolish to expect a ratio of 3 : 1 in each pod. The larger the number of seeds observed the less the error.

The theory accounted for what he had observed, and it proved to have an astonishing power of prediction. It predicted findings that would otherwise be totally unexpected. It will be enough to mention one or two of these.

To continue the experiment with round and wrinkled seeds: the F_2 seeds, both round and wrinkled, were planted out in the following year and the plants allowed to self-pollinate. Mendel's theory predicts that the wrinkled seeds (aa) must breed true; one-third of the round seeds (AA) must breed true; two-thirds of the round seeds (Aa) must behave like the original F_1 hybrids, producing round and wrinkled seeds in the proportion 3 : 1. In his experiments all the wrinkled seeds bred true. Of 565 round seeds, 193 produced plants yielding round seeds only, and

372 produced plants yielding round and wrinkled seeds in the ratio of 3 : 1. That is, the proportion of hybrids to constants in the F_2 round seeds was 372 : 193 or 1·93 : 1—approximately 2 : 1.

Mendel studied the inheritance of 7 contrasting characters in varieties of the garden pea, both singly and in various combinations. The study of combinations of characters made the experiments more complicated and laborious. One of the simplest was the study of a cross of a plant with *round yellow seeds* and a plant with *wrinkled green seeds.*

Once again it will be seen that the law does not predict what is not a fact.

'A' and 'a' having the same significance as before, let 'B' represent the dominant factor for yellow seeds, 'b' that for the recessive green seeds. This cross can therefore be represented:—

	Round Yellow		Wrinkled Green
Parents	AA BB		aa bb
Gametes	(AB)		(ab)
F_1		AB ab	

'A' and 'B' being dominant, all the F_1 seeds should be round and yellow; and this is what was found.

These F_1 hybrid seeds were planted out in the following year and the plants allowed to self-pollinate.

The progeny will clearly depend on whether 'AB' is transmitted as one character, and 'ab' as one character, or whether 'A', 'B', 'a', 'b' are transmitted quite independently.

If the characters segregate independently into the gametes, Mendel's theory predicts:—

Parents	ABab	ABab
Gametes (in equal numbers)	(AB) (Ab) (aB) (ab)	(AB) (aB) (Ab) (ab)
F_2	As shown in the square below	

Female gametes

		AB	Ab	aB	ab
Male gametes	AB	ABAB	ABAb	ABaB	ABab
	Ab	AbAB	AbAb	AbaB	Abab
	aB	aBAB	aBAb	aBaB	aBab
	ab	abAB	abAb	abaB	abab

These 16 combinations are formed in equal numbers. There are 9 types, as shown below, and it is clear that the F_2 seeds must be in the proportion:

9 round and yellow (AABB, 2AABb, 2AaBB, 4AaBb)
3 round and green (2Aabb, AAbb)
3 wrinkled and yellow (aaBB, 2aaBb)
1 wrinkled and green (aabb)

Mendel's figures in this experiment were:—

Of 555 seeds, 315 were round and yellow, 108 were round and green, 101 were wrinkled and yellow, 31 were wrinkled and green. This is a close enough approximation to the proportions 9 : 3 : 3 : 1 which the theory predicts.

A beautifully concise exposition of the experimental evidence for the theory was first read to the Brünn Scientific Society and then published in their *Proceedings*. Either no one in the rest of Europe read the Proceedings of an obscure Moravian society, or those who did were very stupid. No notice whatever was taken of the paper. Mendel became prelate of his monastery, because he was a modest, unassuming man of solid character and high intelligence: it was not because of his brilliant scientific work—no one knew that he had done any. The *law of segregation* was rediscovered 30 years later.

It seems odd to us that anyone could read Mendel's account of his experiments and refuse to admit that his theory was probably true. If we stop to wonder why it seems odd the principle guiding us seems to be that the credibility of a theory increases with the number of its consequences that are shown to be facts. The law of segregation not only explained the facts he had observed; it predicted a great many previously unknown facts.

It is easy to see at least roughly the consequences of most theories, but it may be impossible to calculate them precisely. This is one of the difficulties that may hold up the work of verification. When Newton had thought of the Law of Universal Gravity, he could see that among its consequences would be that the planets moved as they did. But he found it impossible to calculate how his theory required that they should move; the mathematical methods of that time were too elementary to cope with such a task. Before he could make predictions from his law that could be compared with the results of observation, he had to expand mathematics by inventing *calculus*. After this it was possible for him to calculate precisely the consequences regarding the movements of material bodies subject to gravitational interaction. He was able by the use of calculus to predict, for instance, that the planets would be found to move round the sun in ellipses, with the sun in a focus; that the radius vector (the line from the sun to the planet) must sweep equal areas during equal times; that the squares of the time of revolution of the several planets must be in the same ratio as the cubes of their mean distances from the sun. At least Newton did not have the trouble of carrying out observations to see if these predictions agreed with fact. The observations had been made half a century previously by Johannes Kepler.

We have all met people who appear to believe that a theory must be true if its consequences are true. They are, of course, falling into the elementary fallacy mentioned in an earlier chapter, called by Aristotelian logicians the 'fallacy of asserting the consequent'. Where p and q are any propositions, if p implies q and q is true, *it does not follow that* p *is true*. Propositions other than p may also imply q. True, we show that p implies q, q_1, q_2, q_3, . . ., which are all true; but to reiterate a fallacy does not make it valid.

Common sense tells us it is unlikely that more than one theory will imply so many facts. It must be admitted, however, that to accept a theory because we cannot think of another that fits the facts is not a very satisfactory basis for belief.

It comes as a relief to know that certainty can be found in falsification. The one clear decisive thing we may be able to do with a theory is to falsify it. If we find that it predicts what is not a fact, then it is false beyond doubt. For this does follow deductively: *if* p *implies* q *and* q *is false, then* p *is false.*

The evidence that gives credibility to a theory is failure to falsify it. A theory that cannot be proved untrue is to be accepted as true.

It is not always so easy as one might think, however, to show that what a theory predicts is not a fact. It depends to some extent on the kind of prediction.

For Mendel it was comparatively easy. A more or less decisive answer can usually be given when the prediction is numerical. The predictions of his law of segregation, which he tested by experiment, would all have been at once falsified had they been false. Theory predicted, for example, that in the F_2 hybrids dominant and recessive characters would be found in a ratio approximating to 3 : 1 as the size of the sample increased. The experimental result was 2·96 : 1. By a statistical scrutiny* it is possible to state the mathematical probability with which this is 'within the limits of experimental error', i.e., the probability that the deviation from the predicted figure is due to error of sampling. It is high enough to say that the result is what theory requires.

Often what is predicted is that something will, or will not, occur under certain experimental conditions.

It may be quite easy to say decisively whether it occurs or not. For instance, the physiologist, Claude Bernard, in a rash moment, suggested that the fermentation of grapes was the result of a soluble ferment inside the grape and had nothing to do with the vital activities of a living yeast. This was too much for Pasteur. It implies that all grapes will ferment even if the skin of the ripening fruit is protected from the yeast that normally settles on it in late summer, a consequence that should be easy enough to confirm or refute. Though in the middle of important research on anthrax, Pasteur left Paris for a little vineyard he owned at Arbois, the home of his childhood. The grapes were green; it was early August, and on their surface he could find no trace of yeast. He enclosed some of the vines in a glasshouse and wrapped the bunches of grapes in sterile cotton-wool. In October he harvested them. The protected grapes did not ferment; the neighbouring open-air grapes fermented after 36–48 hours, as grapes have always done.

The experimental difficulties, however, can be considerable. Since 1904 there had been two theories to account for the effect of a nerve impulse on muscle supplied by the nerve: (1) The wave of electrical disturbance, which is the nerve impulse, spreads over into the muscle and produces its effect; (2) The nerve impulse produces at the nerve-ending a chemical substance responsible for its effect. Not until 1921

---

* *See* Chapter XVII.

was a crucial experiment devised which falsified the first. This was Loewi's experiment:—

Make a preparation of a frog's heart with its nerves intact, filled with Ringer's solution, and so arrange that the fluid flows continuously to a second frog's heart whose nerves have been removed. If the first hypothesis is true, stimulation of the vagus nerve-fibres to the first heart will cause the typical slowing and arrest of its beat, but will have no effect on the second heart. But what is actually found to occur is that, shortly after arrest of the first heart, the beat of the second heart is slowed and arrested.

It was even more difficult to show the falsity of the first hypothesis in accounting for the contraction of skeletal muscle that follows the motor-nerve impulse.

When a theory predicts that something exists its refutation may be very difficult. Non-observation is not proof of non-existence.

For instance, it has been an accepted theory since the beginning of the century that the cell nucleus is responsible for reproduction: no cell could divide and multiply if it had no nucleus. This implies that there is a nucleus in bacteria. None could be seen in any bacterial cell but this did not falsify the theory. It is only in comparatively recent years, by the help of new staining methods and especially the electron microscope, that nuclear material in bacteria has been demonstrated quite clearly.

There is a saying, often repeated, that 'it is not possible to prove a negative'. I have never known exactly what it means. If it means you cannot prove any sentence with 'not' in it, it is palpably false. I think it may mean 'you cannot observe a negative fact'. If I look out of doors and remark: 'I see it is not snowing', how have I managed to observe 'not-snow'? It will be admitted, however, that this negative judgement from observation is evidence of the same weight as any positive observation, because of the certainty that I would have seen snow had it been there. This is the criterion: for non-observation to have any weight as evidence you must show it to be highly probable that the thing would have been observed had it been present.

It is generally believed that Pasteur disproved once and for all the theory of spontaneous generation. He did nothing of the sort. It would not be possible to observe the non-transformation of lifeless into living matter unless one could be certain of observing such transformation if it did occur. What Pasteur did was to show that the alternative doctrine of homogenesis, *omne vivum ex vivo*, had never been falsified.

Every experienced research worker tries to prove himself wrong, hoping he will not succeed. His critics will try if he does not. He must try to falsify his own theory, though of course he does not want it to be false; he wants to show that it cannot be done.

The way to do this most convincingly is to follow those consequences of the theory that are unexpected, or that appear to clash with common sense. Hence the universal belief that the more unexpected the predictions of a theory, the more likely it is to be true if observation shows that they are facts.

If, for example, you know nothing about heredity, it is very unlikely that you would predict that if you cross a garden pea bearing round yellow seeds with one bearing wrinkled green seeds, the progeny would be all round and yellow seeds; and that these would grow into plants bearing seeds round and yellow, round and green, wrinkled and yellow, and wrinkled and green, in the proportions 9 : 3 : 3 : 1. Mendel's theory gains enormously in credibility by showing that this unlikely prediction is what actually happens.

No one forestalled his critics in this way better than Darwin.

The different response to the work of Darwin and Lamarck has puzzled many people. Both revived a theory well known to the ancient Greeks and to the Renaissance humanists, that all living things on earth are descended from common ancestors. Why was there a contemptuous shrug for Lamarck, for Darwin a torrent of emotional abusive derision, and within ten years acceptance of his theory by all in a position to judge?

Lamarck tried to explain the appearance of new species by a number of unverifiable speculations. The neglect he suffered is the normal response. The prolonged misrepresentation, vilification, and vulgar derision that Darwin had to endure is the normal emotional response to the discovery that Reason contradicts Revelation, and that Reason seems likely to be irrefutable. Darwin had produced a verifiable hypothesis and had carried through the work of verification with matchless care and skill.

The Origin of Species is one of the great books of the world, not only because it fundamentally changed human thought: it is a superb example of the search for truth by scientific method—a long and arduous effort to show that a theory cannot be falsified. Facts beat on the brain, wave after wave: observations that everyone had made or could easily make, to be expected on the doctrine of descent but wholly inexplicable on the alternative doctrine of creation. Darwin produces them one after another in endless variety and with consummate ease. And he applies himself with especial care to those implications of this theory that seem to clash with common sense.

Admitting, for instance, that at first sight it seems absurd in the highest degree that so wonderful a structure as an eye would have come into being through variation and natural selection from a creature with no eye, he describes the almost infinite gradations that can be observed, from lowly creatures with a patch of skin sensitive to light to the true picture-forming eye. At the end of this inquiry he says: 'If it could be demonstrated that any complex organ existed which could not possibly have been formed by numerous, successive, slight modifications, my theory would absolutely break down. But I can find no such case.'

The existence of animal instincts presented him with even more formidable difficulties. 'Many instincts are so wonderful that their development will probably appear to the reader a difficulty sufficient to overthrow my whole theory', he wrote.

An *instinct* is an action performed by an animal, and especially by a very young one, without experience, and performed by many individuals in the same way. The most interesting and baffling of all are the instincts of the sterile and neuter insects. The 'worker ants', for instance, differ not only in structure from male and fertile female ants; they have remarkable instincts not possessed by their parents. They combine to build their houses, keep them clean, and guard them, going into battle in well-ordered array when occasion arises; they go on foraging expeditions to collect food for the community, performing quite remarkable engineering feats in doing so. But these creatures are absolutely sterile and so cannot transmit modifications to progeny; they have no progeny. Their parents do not have these instincts: how then could they have evolved by the preservation of profitable variations of simpler instincts? If you want to know Darwin's answer you must read the *Origin*. It is sufficient to draw attention to the time and thought he lavished on the kind of consequence that at first sight seemed to falsify his theory.

CHAPTER XI

THE LAWS OF NATURE

We may be forgiven if we complain that the light of Reason's lamp, when guiding us to a knowledge of general truths, seems a little dim.

We will concede that by our reasoning faculty we can know of some general propositions that they are probably true.

Some are inferences by *induction*. The principle that would validate the inference is that, if a number of α's have been found to be β's, and none to be not a β, then 'all α's are β's' has a degree of credibility which increases as this number increases. But no one really knows if there is such a principle.

Then there are the theories made known by the *hypothetico-deductive method* which we have just described. We believe that the greater the number of theoretical consequences checked against observation, the greater the chance of falsifying a theory if it is false. Therefore we accept as probably true those theories which have survived prolonged verification of this kind—theories whose predictions have always proved true. Just as, if a system at roulette won all the time without exception, we should begin to think there was something in it.

If this is all that our reason can do it is not very impressive. It is, however, difficult to account for the apparent certainty of some of our beliefs that are undoubtedly based on evidence. We are convinced that if anyone slips off the edge of a cliff he will fall; just to admit that it is probably true would strike us as ridiculous.

But, of course, it is not all. Reason is continually leading us, if not to certainty, to something very close to it.

Imagine that we are back in the early nineteenth century, knowing absolutely nothing about the mechanism of heredity. We have observed on many occasions that if we cross a true-breeding tall garden pea with a dwarf pea, the plants that grow from the hybrid seeds are all tall. On this evidence we think it probable that such a cross will always give us tall plants. As a generalization its probability is not very high: most inductive inferences are wrong.

L_1 may designate laws of this rough and ready kind.

Now imagine that, later on, we have the wit to see that Mendel's law of segregation has a high degree of probability upon the evidence he produced. It will be at once obvious that our first inductive generalization is implied by this law: it is a consequence of Mendel's law that the F_1 hybrids shall be all tall.

L_2 may designate a law which implies an L_1. The evidence for L_1 is much less than that for L_2, which therefore has the greater degree of credibility.

The reader will recall the principle of deduction: If *p* implies *q* and *p* is true, then *q* is true. What is implied by a true proposition is true: if *p* is certain then *q* is certain. He will readily concede that what is implied by a proposition that is probable has the same degree of probability.

It follows that our rather doubtful generalization about crossing tall and dwarf peas suddenly acquired the much higher probability of Mendel's law.

At the beginning of the present century Mendel's law was seen to be implied by the theory of the gene. L_3 may designate this theory, and other laws that imply an L_2. It is obvious that Mendel's law of segregation came to have the same degree of credibility as the theory of the gene. So therefore does any generalization implied by Mendel's law: for if L_3 implies L_2, and L_2 implies L_1, then L_3 implies L_1.

The credibility of an inductive generalization rises steadily as it is seen to be implied by wider and wider laws, the evidence for the more comprehensive law being greater than that for the more restricted law.

I do not think it is known that the theory of the gene is implied by any more comprehensive law of living matter, though probably there is such a law. It is only in physics and chemistry that we can go back to a few laws of very wide generality.

J. S. Mill, over a century ago,* showed the importance of this deductive interrelationship among certain laws. 'If we find any means of deducing one of the less strong inductions from stronger ones', he wrote, 'it acquires, at once, all the strength of those from which it is deduced.'

Any law deducible from a more universal law is not, said Mill, to be called a *law of nature*. 'It is the custom in science', he wrote, 'wherever regularity of any kind can be traced, to call the general proposition which expresses the nature of that regularity, a *law*. But the expression *law of nature* has generally been employed with a sort of tacit reference to the original sense of the word *law*, namely, the expression of the will of a superior. When, therefore, it appeared that any of the uniformities which were observed in nature would result spontaneously from certain other uniformities, no separate act of creative will being supposed necessary for the production of the derivative uniformities, these have not usually been spoken of as laws of nature.' Few people, and no scientists, now use the word *law*, in *law of nature*, with the connotation of an edict. Like any other scientific law, a law of nature means nothing more than a uniformity to be found in nature.

According to Mill the *laws of nature* are the fewest general propositions from which all the uniformities which exist in the universe might be deductively inferred.

The trouble with this definition of *laws of nature* is that we have no idea what are these fewest general propositions. The *law of universal gravity* may be one; for, if Einstein's view is correct, it is implied by the structure of the universe. But no one knows what is the connexion between gravitational and electromagnetic forces, and it may be that this connexion will be revealed by some simpler law than the law of gravity. In the end it may be shown that there is only one *law of nature*, from which all the rest can be deduced.

---

* MILL, J. S. (1843), *System of Logic*, Book 3, Chapter VI. London: Parker.

What is necessary at the moment is to draw attention to those universal laws—call them *laws of nature* if you will—which are not obviously implied by simpler laws, and from which a large number of the generalizations of a science follow as consequences. Thus the law of universal gravity is a law of nature in this sense, and from it can be deduced the laws of planetary motion, the laws of the tides, the laws of falling bodies, of projectiles, of the curious behaviour of spinning bodies, and a host of minor generalizations. A system of laws so related is called a 'deductive system'.

The evidence that raises the probability of a general proposition to something approaching certainty is the evidence that shows it to belong to such a deductive system: for when it is shown to be ultimately deducible from a law of nature it at once acquires all the high probability of that law. The probability of any generalization that could be deduced from the law of universal gravity (say, Galileo's laws of falling bodies) is that of the law of gravity itself. To deny it would be to deny the whole deductive system. To deny any one member of the system would be to deny all. We do not believe that if we slip off the edge of a cliff we shall fall because so far everyone who has slipped off has fallen: we know that 'all unsupported bodies fall' can be deduced from the law of gravity and therefore has the same certainty.

All the theorems of Euclid are true *if* the axioms from which the reasoning starts are true. Similarly, all the general propositions of a scientific deductive system are true *if* the law of nature which is their source is true.

For the problems of applied science, and those of day-to-day life, we accept that laws such as the law of universal gravity, the laws of electrodynamics, the atomic theory, the theory of the gene, the principle of evolution, and so on, are so near the truth that they can be regarded as certain. This judgement is justified by the following consideration.

As the decades, or centuries, have gone by, more and more general propositions, at first seen to be probably true on their own evidence, have been found to be consequences of one of these ultimate laws. In sciences like physics and chemistry their number is now enormous; they make up the whole corpus of knowledge of these sciences. The probability that by this time the law of which they are consequences would have been falsified, had it been false, is so near certainty that the difference is negligible. And this, of course, is the probability that the law is true. Every new induction discovered to be deductively related to other laws of the science, not only acquires the strength of the law of nature which is the source of them all, but strengthens it.

But, a philosopher would object, there is this fundamental difference between the theorems that form the deductive system of mathematical logic and the general propositions that form the deductive systems of physics and chemistry: the former follow deductively from the axioms of logic which we cannot doubt without refusing to think at all; the latter follow from a few *laws of nature* which cannot be proved and which we are not compelled to accept. We cannot, therefore, know that any of the general propositions of natural science are true as we can know the truth of the theorems of logic and mathematics. It is not so very long ago that the law of the conservation of matter was considered certain.

CHAPTER XII

THE PURPOSE OF THINGS

CURIOUS people who ask questions beginning 'Why?' want to be told the cause of a thing or its purpose.

When the question concerns living things or human activities they can often be given both. Why do spiders spin a web? Because an inherited instinct makes them do it, to catch flies. Why does the heart beat faster when we are frightened? Because adrenaline is poured into the blood-stream, so that we can fight better or run away more quickly. Why do trains go from London to Brighton? Because they are driven, by means of electric motors, to take people who want to go there.

But when the question is about something in the inanimate world there is only one answer. Why does the earth move round the sun? We must be content, modern science affirms, with knowing the cause: because it obeys the laws of planetary motion. The Greeks of classical times would have asked: 'For what purpose?'

In the prescientific era the purpose of a thing was held to be the really important 'cause' of its existence.

The concept of a 'cause' is a peculiarly difficult one. If you ask: 'What do you mean by the cause of an event?' most people will answer: 'What makes it happen.' But this is a vague and rather muddle-headed notion. What kind of thing makes an event happen?

We ourselves can do it: in our daily lives we are continually making things happen. But clearly a vastly greater number of events happen with which we have nothing to do. Well then, the Greeks of Aristotle's time would have argued, something must have done it: and what more likely than beings like ourselves, with the same sort of purposes? It is the gods, of course: if not one of the Olympian gods, then Fortune, or Fate, or the Sun, the Moon, or one of the Planets.

Pervading the whole of Greek thought was a notion that everything was made to happen for some purpose. It is, they would have said, easy to see what some of these purposes are. Rain is sent to help growing crops; the wind to carry our ships along; the sea to nourish the fish we need for food; storm and calm, drought, and all baneful things, to punish us for some error.

This was the background of thought in Aristotle's time. With his acute logical mind he saw that it was still possible to inquire how the gods do it—to reveal the mechanism by which events come about. Rain may be sent by Zeus to help growing crops: it is still possible to say that rising exhalations cool, and, cooling, become water which sinks.

Aristotle distinguished four 'causes': four parts of what is responsible for the existence of a thing or the occurrence of an event. A thing is explained, he taught, when you know these causes.

By far the most important of the four was the *final cause*: *the end for the sake of which* the thing exists or the event occurs. It seems odd to us to call the purpose of an event its 'cause'. But by 'cause' Aristotle meant 'explanation', or 'making clear'; and a thing is certainly much clearer to us if we know what it is for. Seeing the purpose of the valves in the veins certainly helped Harvey to discover the circulation of the blood. Things are explained by the purpose they serve, and of course the agent which foresees and attains it. This for Aristotle was *φύσις*, which we translate 'Nature'. The term 'final cause', he says, has a two-fold signification, as it implies that for which as well as that by which any result is obtained.*

But though, as Aristotle said, 'Nature does nothing without a purpose', that purpose must be achieved by some means or other. And the means are analysed into three causes: (1) The *efficient cause*, which sets a process going; (2) The *material cause*, the matter in which the process evolves; (3) The *formal cause*, which moulds the formative process until the end is achieved. The 'efficient cause' is very close to the modern connotation of 'cause': it tells *how* the purpose is achieved. The three are what we would call a *mechanistic explanation.*

The doctrine that an event takes place because it serves some purpose is called 'teleology' (*τέλος*=end); and explanation by final causes is a *teleological explanation.*

'I do not see how it could have been known in advance', Bertrand Russell has written, 'which of these two questions science ought to ask, or whether it ought to ask both.'

Aristotle's doctrine of final causes persisted, though much distorted, throughout the Middle Ages. For the medieval clerics purpose was the only cause that need be considered; but they were thinking of divine purpose. God's purposes were a sufficient explanation of anything. And they were ubiquitously obvious in the world; a world with its varied living things so clearly provided for the needs of man.

Speculation about the means whereby the divine purposes were achieved bordered on impiety. Any interest in the world around us, except in so far as it manifested the wisdom of God, was frowned on. Any curiosity to know more than this about it was firmly suppressed. It was in any case a waste of time, for this world is of little or no importance compared with the world beyond the grave.

It took more than a century for the Renaissance humanists to break down the medieval Church's opposition to the mechanistic question. But curiosity about the means by which purposes were achieved could not be suppressed indefinitely. Gradually the pioneers of modern science began to ask for 'causes' in the modern sense.

A modern scientist would say that in the lifeless world there is no evidence of any purpose in events. He can see nothing at all comparable to the activities of living things. It is not, he would say, a self-evident truth, as the old Greeks seemed to think, that the planets swing round the sun for a purpose. In the inanimate world, therefore, questions about purpose seem to be inapplicable, and as a scientist he is compelled to limit his inquiries to description of what happens and how things happen.

---

* ARISTOTLE (1855), *On the Vital Principle* (trans. COLLIER, C.). London: Macmillan.

Students of the physical and chemical sciences do not assert, as a number of immature critics have implied, that what they cannot see does not exist. It is true that they look at only one facet of a problem, but they do it deliberately. They have chosen to ask only the mechanistic question.

Most people would agree that questions about purpose are rightly excluded from the physical sciences. But things seem to be very different in the living world. Aristotle was the first man in the world to think seriously about living things, and naturally he saw purposive behaviour everywhere. Although he completely rejected mythology and the Olympian gods, he believed that there was divine guidance of the universe, and that the ultimate purpose of living things might be some transcendent goal. But the final causes he looked for and talked about were much less visionary. The final cause of a developing chick was the full-grown hen; the final cause of an eye was vision.

All animal and plant behaviour is purposive. It is, in fact, purposive behaviour which distinguishes living from lifeless matter. Should not the biological sciences, therefore, ask both mechanistic and teleological questions?

Until we think more carefully about it, 'purpose' seems to imply conscious mental activity. We are so familiar with human purposes. We consciously resolve on some end which is clearly foreseen. Sometimes, after a deliberate choice from alternative means, we plan activities designed to attain it, and we call the desired end our 'purpose'. But a little reflection reminds us that many of our actions that we would call 'purposive' are not preceded by any such conscious mental activity. If we see a child straying across a busy road we act impulsively, without any reflection at all, unconscious of having any purpose, and snatch it out of danger. And we do hundreds of purposive things by habit: we will walk on a crowded pavement and quite unconsciously take steps to avoid other pedestrians.

Then again, psychologists are agreed that our activities may be controlled by wholly unconscious purposes. A sufferer from hysteria, or another neurosis, may engage in purposive activities while wholly unconscious of their purpose.

It would seem then that 'purpose' does not imply consciousness. But does it imply mind?

Let us forget human resolutions and strivings for a moment and look at purposive behaviour in lowly living creatures.

Aristotle relates in *Historia Animalium*, the first book on natural history ever written, the story of the *ichneumon* wasps. These wasps kill spiders and carry off their dead bodies to a hole in a wall or some such place. On the dead spider they lay their grubs, then smear over the hole with mud. From the grubs come the young wasps, and there to hand is the food they need. This is the sort of thing that would occur to a wasp, we might think, foreseeing the needs of its future progeny.

Every kind of instinctive behaviour benefits the organism or its species. When we consider the intricate and delicate manœuvres of a spider spinning its silken web, then waiting in an obscure corner with one leg on a strand of it, poised to detect the vibration made by its insect prey, it is hard to believe that this behaviour is not directed by something that has an eye on what will happen.

But we do not attribute conscious planning to the wasp or the spider: it is widely agreed that we need not assume that any instinctive behaviour is associated with conscious purposiveness. Must we attribute mental activity at all to them? A critic, influenced perhaps by the word 'purpose' which strongly suggests mental activity, may say, 'Why not?' Why not admit that purposive behaviour is a manifestation of mind?

It may be that, in view of the extremely close relation of mind and brain in human beings and higher mammals, we must predicate 'mind' of any creature with a brain, however primitive. But here is a remarkable instance of purposive behaviour in an invertebrate with nothing that could reasonably be called a brain.

Microstoma is a minute aquatic flatworm which attacks, or defends itself, by discharging nematocysts or stinging cells from its surface. These weapons once belonged to another microscopic animal, a freshwater *Hydra*, from which *Microstoma* has captured them. What happens is this. *Microstoma* eats *Hydra* and digests all except the nematocysts, with which its tentacles are heavily armed. These stinging cells are passed through the worm's enteron and are then taken up by wandering tissue cells and carried to the ectoderm. Here they are positioned uniformly over the surface and with the poison tube directed outwards. *Microstoma* eats *Hydra* for its nematocysts: when its stock of weapons is full it will, even when hungry, refuse the animal as food.

And in all animals purposive behaviour is seen not only in the activities of the individual living creature; every organ in its body behaves in a way helpful to the creature as a whole. Consider, from a host of similar automatic reflex actions, the changes that have been shown to take place in the body of a cat when suddenly faced with a dangerous situation. There are two courses open to it: it can fight or run away. In either case the most strenuous muscular exertion it is capable of will be needed. So the heart beats faster and the blood-vessels of the muscles dilate, whereas those of the skin and internal organs contract. The effect is a prompt and considerable increase in the volume of blood going to the muscles, and the minimum going where it is not especially needed. At the same time the spleen contracts and pours a reserve of red blood-cells into the circulation, ensuring a good supply of oxygen to the muscles. And again, the amount of sugar in the blood is promptly increased, in order that the muscles shall have the food energy they are going to need.

It is not only the organs of the animal's body that behave in this altruistic way; every one of its cells does so. Consider their response to a simple tissue injury. Local capillaries dilate, and through their walls certain white blood-cells emigrate into the dead tissue to begin the process of its removal. Quickly they are joined by considerable numbers of other cells to carry on the work, and with new capillary vessels which grow into the injured area to carry out the process of repair. The chain of events comes to an end with replacement of dead tissue by scar tissue: every cell has behaved as if it knew what to do, and the purpose of it.

And so, if we must attribute mind to the wasp and the spider we must do so to lowlier creatures still: to every animal and plant cell, to every unicellular animal like an amoeba, to every bacillus, and to every virus particle.

It must be admitted that organic activities tend to a goal which benefits the organism or its species. This is something observable, not found in the inorganic world, but peculiar to living things.

The idea, therefore, that things and events are, at least in part, explained by their purpose does seem to be applicable in the living world; and biologists have continually asked to be enlightened about the purpose of things. There must be few animal or plant tissues whose function is not now known, or instances of behaviour whose object is in doubt. Usually the answer has been fairly obvious. It does not take much observation to see why a spider spins a web. Often it has been far from easy. Before 1856, for instance, no one knew the function of those little flattened triangular bodies shaped like a cocked hat, one above each kidney—the adrenal glands. In 1716 the Bordeaux Académie des Sciences had offered a prize for an answer to the question: 'Quel est l'usage des glandes surrénales?' It could not be awarded.

But an answer to a simple question: 'What is the purpose of this?' which gives only the goal, and omits all reference to what directs things towards it, is not what is meant by a teleological explanation. There is nothing here to make biologists bristle with indignation, as they do if accused of teleological thinking. Teleology is not identified with inquiries into the biological function of things, but with Aristotle's suggestion that living matter differs from not-living matter in the possession of a mysterious directing vital principle, and especially with the emergent *vitalism* of more recent times.

The conflict is between a mechanistic and a teleological explanation of so-called *purposive behaviour*.

Aristotle was a thorough-going teleologist. One of his major contributions to biology was a little book called Περὶ Ψυχῆς. It is translated into Latin: *De Anima*, and into English very often as: *On the Soul*, though much better as: *On the Vital Principle*.

In this book he tries to explain what it is that makes living matter different from the stuff of the rocks. It is not an easy book: he is tackling the most fundamental problem in all biology. But those who translate ψυχή as 'soul', it seems to me, misunderstand his thought. It is some indication of the change in meaning of 'psyche' in the centuries that have passed since he selected it to denote that which distinguishes living from not-living matter. For it did not mean anything like what 'soul' means to us, whatever that may be. For Aristotle it was a part of living matter responsible for its life, and not separable from it. It was not something that is capable of surviving the living body. Living matter was matter endowed with *vital principle*.

Vital principle is the final cause of life: 'If the eye were an animal vision would be its vital principle.' It is Nature (φύσις) in matter and at work in matter, achieving the purposes of life, which can be summarized as reproduction, sensibility, and reason. It can therefore be analysed into a *vegetative principle* common to all living things; an *animal principle*, in animals and man, but not in plants; and a *rational principle*, in man only.

Whatever kind of living matter it is—plant, animal, or human—Nature is at work in it, like an artist or a carpenter, using it for the purposes of life. In this, and in this alone, it differs from the matter of the earth and the rocks.

This view clearly has something in common with the *vitalism* of comparatively modern times. It will be recalled by those who know a little of the history of biology that this doctrine persisted until close on one hundred years ago. *Vitalists* believed that living matter was governed in its behaviour by special laws of its own. They were not the laws of chemistry and physics. Presumably they were unknowable by the human mind. Accordingly, any attempt to interpret the changes that occur in the animal body by the laws of chemistry and physics was doomed to failure: laboratory investigations of living creatures were a sheer waste of time.

How vague their ideas were can be seen from the comments of several eminent biologists at the end of the eighteenth and in the early nineteenth century. Thus, John Hunter (1728–93), a surgeon who devoted his life to comparative anatomy and physiology:—

> Animal matter is endowed with a principle called, in common language, *life*. This principle is perhaps conceived of with more difficulty than any other in Nature, which arises from its being more complex in its effects than any other; and it is therefore no wonder that it is the least understood. But, although life may appear compounded in its effects in a complicated animal like man, it is as simple in him as in the most simple animal, and is reducible to one single property in every animal. . . . The first and most simple idea of life is its being the principle of self-preservation, by its preventing matter from falling into dissolution—for dissolution immediately takes place when matter is deprived of it; the second is its being the principle of action. These are two very different properties, though they arise from the same principle.

Again, John Barclay (1758–1826), a leading teacher of anatomy:—

> In every living organized structure there is plainly a power that preserves, regulates, and controls the whole; directing at first the different processes in forming one part of the organs, afterwards employing the assistance of the organs which it has formed to produce more, till at last it completes the whole of the system in such a manner as to suit its future conveniences and wants. This power, or rather this agent, physiologists have named *vital principle*. . . . But in all operations that are performed without either volition or consciousness, it appears subordinate to a much higher power—to the Almighty and Omniscient Being who dispenses his laws to the boundless universe, and whose laws, except by himself, can never be improved, altered, or abrogated.

And Marie François Xavier Bichat (1771–1802) produced this lovely example of *circulus in definiendo*: 'The *vital principle* is the assemblage of the functions which resist *death*.'*

It is true that later on some biologists compromised. Johannes Müller (1801–58), the great German physiologist, suggested a compromise widely accepted, i.e., that there are two sorts of living phenomena: one the result of physico-chemical causes and amenable to experimental inquiry, the other explainable only by the action of a vital principle and outside the domain of science.

---

* Quoted by COLLIER, C. (1855), in *Aristotle on the Vital Principle*. London: Macmillan.

The vitalists were opposed and finally defeated by the *mechanists*, who believed that every event in living matter would ultimately be explainable by the laws that govern events in the inorganic world.

Most modern biologists, though not quite all, would say that biological behaviour can be explained without assuming any non-material directing agent. They have no need of any such entity. The fourteenth-century philosopher William of Occam is reputed to have said—a maxim since known as 'Occam's razor'—*entia non sunt multiplicanda praeter necessitatem* ('entities are not to be multiplied without necessity'). In other words, if explanation is possible without assuming some hypothetical entity, do not assume it.

Purpose, the modern mechanist would say, implies neither consciousness nor mind. The impression we get that the purposive activities of lowly living things imply a guiding intelligent mind, foreseeing an end, resolving on it, and striving after it, is pure illusion. The wasp and the spider behave as they do because they can do no other. Some of their gene patterns, which they have acquired by millions and millions of years of evolution, are a code of instruction for this behaviour, which they must follow as the planets must swing round the sun.

It is true that we often find it convenient to personify Nature, as if she were designer and craftsman. Only the other day I heard an experienced naturalist say 'the albatross's wings are *designed* for gliding'! But this is purely figurative speech. Looking at the progress in the biological sciences during the past century we are almost irresistibly brought to Bertrand Russell's conclusion: 'Experience has shown that the mechanistic question leads to scientific knowledge, while the teleological question does not.'

CHAPTER XIII

THE NOTION OF A CAUSE

If you ask someone engaged in cancer research, 'How do you know that cancer has a cause?' he will say that there are no uncaused events. If you press him and ask how he knows this, he will either ask you to stop wasting his time, or say simply that the *law of universal causation* is the faith by which scientists live.

Aristotle would have given you the same look of polite pity if you had suggested to him that some things happen without any purpose.

Can a scientist give better reasons for his belief that everything has a cause than the old Greeks could give for their conviction that everything has a purpose?

It is important to consider first, not why scientists believe the law of causation, but exactly what it is that they believe.

Having observed that the course of Nature is uniform and not chaotic in that part of the universe with which we are acquainted, we have come to have an unshakeable belief in universal uniformity. A falling body falls with constant acceleration and not unpredictably: it always does, whenever the observation is made, and we conclude that throughout the whole unobservable past it has always fallen so, and will continue to do so throughout the limitless future. Copper conducts electricity, and sugar dissolves in water, not only on Monday, Wednesday, and Friday, but at all times. We believe, in the words of J. S. Mill, 'that what happens once will, under a sufficient degree of similarity of circumstances, happen again, and not only again, but as often as the same circumstances recur'.

Of the many thousands of simple uniformities observable in Nature, it will be seen that some are *uniformities of sequence*: an event of a class A is always followed by an event of a class B. For instance: Ignition of a mixture of oxygen and hydrogen is followed by an explosion; Occlusion of an end-artery is followed by tissue death; Increased pressure on a gas is followed by reduction in its volume; Night is followed by day. Other observations are that something of a certain kind, A, is always associated with something of a kind, B. They are *uniformities of simultaneity, or co-existence*: Air has weight; Copper conducts electricity; All living things reproduce.

For the moment let us consider only the uniformities of sequence: these are the key to the popular notion of 'cause'.

'The *law of causation*', Mill says, 'is but the familiar truth, that invariability of succession is found by observation to obtain between every fact in nature, and some other fact which has preceded it.' We call the invariable antecedent the 'cause', and the invariable consequent the 'effect'.

Is the relation of cause and effect no more than invariable sequence? There is something wrong. Day invariably follows night, but we do not say that night is the cause of day.

In defining the *cause* of anything as 'the antecedent which it invariably follows', Mill makes two things very clear.

First, he makes it clear that he does not mean 'which it invariably has followed in our experience': he means 'invariable as long as the ultimate laws of nature hold'. Night, he points out, would not be followed by day under all imaginable circumstances: the sun might cease to rise even though the laws of matter remain as they are.

Night, therefore, is not the cause of day. 'Causation' is 'invariable sequence' only if we understand that the sequence, besides being invariable, is also *necessary*.

The idea of *necessity* seems to be an important ingredient in the notion of cause and effect. Most people think of a cause as in some way compelling the effect to occur: the analogy is with a human volition, a cause making something happen much as we can do. But this will not do—it is a primitive anthropomorphic notion. A billiard ball striking another necessarily produces an effect, but it has no intention of doing so, nor purpose in doing it. When we say that B necessarily follows A, we do not mean that in some mysterious way A *compels* the occurrence of B: we mean simply that B will follow A whatever supposition we may make about all other things.

Removal of all support from a body in the earth's gravitational field is invariably followed by the body falling towards the earth. This succession is invariable under all imaginable circumstances, so long as the law of gravitation holds. We can therefore say that removal of support is the cause of the body falling.

There is rarely if ever invariable sequence between an event and a single antecedent event: it is nearly always a concurrence of several antecedents which is necessarily followed by the event. We may, in popular speech, single out one of these and call it the 'cause': but the real cause is the whole set of antecedent conditions. Mill's second qualification of 'invariable sequence' is demanded by the obvious fact that no sequence is invariable unless there are no preventing or counteracting causes among the antecedents.

There must always be an interval of time between the cause and the effect, and in this interval something might happen to prevent the effect. A bite by a rabid animal is the cause of rabies. The interval between the bite and the onset of the disease is 4 weeks or longer: if, during this interval, a non-immune victim is artificially immunized, he will not develop rabies.

The interval between cause and effect must sometimes be exceedingly brief, but there must always be one. It is hard to see what could prevent an explosion following the occurrence of a spark in a mixture of oxygen and hydrogen, but perhaps an omnipotent demon could insulate the spark before the reaction began.

The definition of the *cause* of anything must therefore be: the antecedents which it invariably and necessarily (or unconditionally) follows, provided there are no counteracting conditions among them. As Bertrand

Russell so characteristically puts it, 'a law A causes B means "A will cause B unless it doesn't"'.*

In the 'cause and effect' with which the *law of universal causation* is concerned, there is a reciprocal implication: the cause implies the effect, and the effect implies the cause. In repeated instances the same cause must produce the same effect, and the same effect must have been preceded by the same cause. The law of causation means that every event has necessarily followed an antecedent set of conditions, such that every instance of the set would be followed by the same event, and every instance of the event would have been preceded by the same conditions.

But in popular speech we use the word 'cause' much less precisely.

We would all agree that, however often the repetition, the same cause must produce the same effect. Whenever anyone strikes a match in a gas-filled room we expect an explosion. If the same cause need not produce the same effect, an explosion would be no more likely than a shower of rain. Life would be peculiarly difficult.

That different instances of the same effect must have the same cause, however, seems contradicted by experience. We would say that there are many possible causes for most events. If a fire breaks out in your house you want to know which among several possibilities was the cause of it. And so for almost any unexpected event. In medicine the plurality of causes is very evident. Sciatica, we say, has many causes: it may be lumbar-disk protrusion, carcinoma of a vertebra, a pelvic growth, and so on.

The explanation lies in the ambiguity or vagueness of the word 'cause', when used as a word of common speech. The antecedent that is invariably followed by the consequent is almost always a group of conditions. Ambiguity usually arises because we are very apt to select one only of these conditions, especially the most obvious, and call it the 'cause'. Suppose the group of conditions necessarily followed by a given effect is A, B, C, and D, and suppose all are present except A. The moment A is introduced the effect follows, and, perhaps naturally, we call A the cause of it.

Gas is escaping into a room, the windows are closed, a certain proportion of gas and air develops: nothing happens until someone walks in and strikes a match. Then we say that striking the match caused the explosion. It did not: it was only a 'causal factor'.

We say that a specific micro-organism is the cause of a disease. It is not. There is an illuminating story about Max von Pettenkofer, a pioneer of hygiene and public health in Germany. He rejected the claim that the cause of cholera was Koch's cholera bacillus. He asserted that environmental conditions, and special conditions in the body, were essential for infection. It is said that he obtained a culture of the cholera bacillus from a fatal case of the epidemic of 1892, and drank a long draught of it. Nothing happened. The causal conditions of an infective disease are more than the specific micro-organism: they include invasion of tissues, and in the tissues a whole group of conditions that we sum up as lack of resistance to infection.

---

* RUSSELL, B. (1948), *Human Knowledge*. London: Allen and Unwin.

The meaning of 'cause' is very commonly restricted in this way to something less than the set of circumstances needed for the effect to follow.

What we have been doing, and still do, in immature sciences like biology, in applied science, and in the business of everyday life—and it is clearly a very useful thing to do—is to take the earlier part of a process and call it the 'cause' of what follows. It does not matter how much or how little of the earlier part of the process we call the 'cause' of the rest, if it serves our purpose.

For instance, a social worker, interested only in prevention, might say that the cause of a man's tuberculosis was the infection of his wife; a bacteriologist would select a slightly later stage and say it was inhalation of tubercle bacilli; an immunologist would say that it was low resistance that made possible successful invasion of his lungs by the tubercle bacillus; a physician might give later stages of lung destruction as the cause of his symptoms; a pathologist, whose sole concern is to elucidate the whole complex evolving pattern of events that we call a disease, can describe the process without using the word 'cause'.

In general terms, a causal sequence is expressed by a general proposition of the form: 'For all x's, x is A implies x is A$'$, where A and A$'$ are the earlier and later stages respectively of some process.'

A 'cause' is an artificial notion retained because of its practical utility: it enables us to make things happen, or prevent their happening. But in our efforts to understand the world around us it is a vague and useless notion. In the physical sciences the word 'cause' need never be used. We must agree with Bertrand Russell's comment: 'The reason why physics has ceased to look for causes is that, in fact, there are no such things.'* It makes for clarity to define the law of causation without the word, rather than by such expressions as: 'Everything that has a beginning has a cause.'

There are no entities called 'causes', with the power of compelling events to occur. All we know is that everything that has a beginning, in following certain antecedents exemplifies some sequence which we have observed again and again, and so think is probably invariable in the sense already indicated. We avoid all ambiguity and woolly thought if we use the word 'cause' only with some practical end in view. Otherwise we should think of *causal laws*.

The faith of a scientist is in the *uniformity of nature*. He believes that the universe is governed by a few general laws, and that everything that happens in the universe happens in accordance with one or more of the innumerable uniformities that we know by observation, or that flow as consequences from these general laws.

Such uniformities are not only uniformities of sequence. Uniformities of co-existence, such as 'air has weight', are also 'causal laws'; so are uniformities of persistence, such as the first law of motion. All are to be called *causal laws*, for all are expressed by general propositions whose form is: 'For all x's, if x is an α, then x is a β. That there is or is not a

* Russell, B. (1950), 'On the Notion of Cause', in *Mysticism and Logic*, Chapter IX. London: Allen and Unwin.

time interval between being an α and being a β makes no difference to the fact of implication, which is the essential character.

It is because we know such laws that we can make predictions. Bertrand Russell has defined a 'causal law' as 'any general proposition in virtue of which it is possible to infer the existence of one thing or event from the existence of another, or of a number of others'.*

When we say that an event occurs in accordance with a causal law, we mean that it exemplifies this uniformity. We are merely stating that it does occur in this way—a statement quite free from any notion of a compelling cause.

Can a scientist give any reasons for the faith that is in him?

How do we know that everything happens in accordance with causal laws; and how do we know that causal laws are the same now as they have been in the past and will continue to be in the limitless future?

Mill held that it is an induction from experience. It has been found in innumerable instances that an event happens in accordance with some causal law, and we know of no exceptions.† We conclude that everything happens in accordance with causal laws. We have found that laws have held unchanged throughout our experience; we conclude that all laws are unchangeable. The argument is induction by simple enumeration. Though it seems weak evidence for so strong a belief, this is probably its only logical justification.

We could consider the consequence of denying it. Let us assume that our belief in the uniformity of nature is false. Suppose first that the contrary is true—the extreme position that there are no causal laws. It is false, for all phenomena, that 'what happens once will, under a sufficient degree of similarity of circumstances, happen again'. A fire that warms you one day may chill you the next. Nature is not uniform, but chaotic and fortuitous. It follows that we can never reason from 'some' to 'all': that all generalizations are false, and that if any prediction is true, it is only by the caprice of the goddess Fortuna. Now we know by experience that this is not so. An assumption which implies that it is must therefore be false.

But to deny the uniformity of nature is not necessarily to assert the contrary. It is enough to assert the contradictory: that *some* events are uncaused, or that *some* causal laws are not invariable.

In the physical sciences many minor generalizations can be shown to be implied by some law of wider generality; and this in turn to be implied by a law still more general. If one generalization is false then so are the laws by which it is implied—up to one of the ultimate laws of the universe. Thus, if it is false that a freely falling body falls with constant acceleration, then the law of gravitation is false. Physical laws are so deductively

---

* RUSSELL, B. (1949), *Our Knowledge of the External World*, Lecture 8: 'On the Notion of Cause', London: Allen and Unwin.

† There is one exception. It seems that quantum changes and radio-active disintegrations in single atoms have no invariable antecedents, and occur in accordance with no law. But it is held that causal laws in 'macro-systems' are statistical laws; the law of causation is not falsified by events in sub-atomic physics.

related that, if one is shown to be false, many others fall with it. If many are false the whole system of physical knowledge is in danger of collapse.

It can be seen that, at least in the physical sciences, this assumption has almost the same consequence as the assumption that there are no causal laws at all: that all scientific knowledge we think we have is illusion; that all general propositions are false, and all predictions invalid. And this we unhesitatingly reject.

A proposition must be true if its contradictory implies what is known to be false. We believe, therefore, that the law of causation, or the uniformity of nature, has a high degree of credibility. Though it is not known to be true, it is clear that a scientist can give better reasons for his faith than could the ancient Greeks, or the medieval teleologists, for their belief that everything has a purpose. It is less clear that he understands what it is that he believes.

CHAPTER XIV

OBSERVATION AND ERRORS OF OBSERVATION

THREE ingredients go to the making of a good observer: two acquired, one innate.

The wide-awake 'observant' person, who, in everyday life, notices everything, often observes indiscriminately and pointlessly, not only the unusual and odd, but whatever it may be that comes within his range. Asked to describe someone who stopped us a few minutes ago to ask the way, most of us might with an effort be able to recall that it was a man; our observant friends will at once give his approximate age and height, say whether he was of slender or stocky build, and describe his clothes and accent.

But observing with an inquiring mind is a rather different occupation from this pointless automatic recording. Nor is the trained power of observation, which contributes so much to successful research, the capacity to stand wide-eyed and open-mouthed in the face of unfamiliar facts. It was not well understood by some of our predecessors that a science is not an accumulation of observed facts, nor does it progress simply by their collection. Magendie, whose name is well known to physiology students, was one who made this mistake: 'I go about the streets of science collecting what I find', he said, comparing himself to a street scavenger. Claude Bernard was his pupil, and succeeded to his Chair at the Collège de France in 1855. The methods of master and pupil are in the sharpest possible contrast. As Michael Foster shows, in his life of Claude Bernard, compared with Magendie's ceaseless unguided observations Bernard's were few, but they were guided by hypotheses and were incomparably more fruitful. Scientific inquiries are often started by a single observation, at most a very few, of the spontaneous kind. But all those that are made for the purpose of the inquiry are planned to test some theory. 'How odd it is', Darwin wrote, 'that anyone should not see that all observation must be for or against some view if it is to be of any service.'

It was the nineteenth-century astronomer, Sir John Herschel, who first suggested the appropriate names *passive and active observation* to distinguish these two ways of looking for facts.

Sustained curiosity is the well-spring of active observation. Curiosity is an innate propensity aroused by something we do not understand, and satisfied by what we consider to be 'explanation'. Like an instinct it is only satisfied when its end is attained. It is not, as a rule, aroused by the familiar, or by what can at once be compared with something familiar. You must first see that there is a problem. Few people are intensely curious to know what the hidden face of the moon looks like, but if the moon began to dance in the sky everyone would be curious. It is the element of surprise that stirs our curiosity. No one, I imagine, would be very interested by an observation that there are 30,000 pebbles under each square yard of Brighton beach. But when Leeuwenhoek, in the seventeenth century, reported to the Royal Society that calculation from his microscopic examinations had shown that there are 30,000 living creatures in each tiny drop of ordinary water no bigger than a millet seed, he aroused intense curiosity throughout the scientific world.

The only serious obstacle to becoming a good scientific observer is a curiosity that is strangely evanescent. What as a rule is lacking is the ability to see a problem. It is odd how many people see no problems anywhere, believing that they already understand most things. Ask an average person if it is not strange that two parents should generate a new individual so very like themselves, and as likely as not he will say: 'Why shouldn't they?'

The people who display an avid and sustained curiosity about some fact they have noticed are those who know they understand nothing to speak of. This propensity is an essential ingredient in the making of a good scientific observer. You will see it intense and insatiable in the Greeks of classical times, in philosphers and pure scientists, and in many children. I doubt if it can be acquired by training.

Any student of the medical or biological sciences who discovers that he has not been endowed with an undue inquisitiveness would do well to specialize in a purely descriptive science such as anatomy, or histology, or natural history; or in some technical skill such as radiology or clinical pathology. Certainly he should not consider a career in research or even in clinical medicine.

Anyone well endowed with curiosity can excel in scientific observation if he is prepared to know enough of his science to gain facility in active observation, and if he is prepared to devote time and patience to the acquisition of technique.

The kind of view for or against which most scientific observations are made may be, like that which sustained Karl von Baer in his laborious studies, the supposition that with persistence certain regularities will be revealed. It may be a hypothesis like Darwin's theory of descent, demanding ceaseless active observations in the arduous work of verification. Usually it is some supposition made on the spur of the moment, and often as quickly rejected, to explain what strikes us as odd.

Automatically and almost subconsciously a trained observer sees a number of possible explanations for what excites his curiosity; and he sees a number of facts verifiable by a more careful and extended observation, which any one of these explanations would entail. He promptly looks for such facts. Almost from the beginning his observations become guided by hypotheses. They are now active observations; they might be called 'experimental', for an experiment is no more than a deliberate looking for what a supposition leads you to expect. A trained observer knows what to look for and what to ignore; in consequence he sees not necessarily more, but more that is important than an untrained person who is merely observant.

Trained scientists differ considerably among themselves in their powers of observation. Some doctors excel at clinical observation: direct observation of ill people, aided at most by the simplest instruments. It is generally agreed that physicians of the nineteenth century were better at it than anyone we could produce today. What is the difference between a good clinical observer and one who is not so good? The doctor who excels overlooks nothing, whether positive or negative, that is relevant, and records nothing that is irrelevant.

A student at the beginning of his work sees just an ill human body and feels bewildered. He is taught to make a routine, in every patient, of examining the systems of the body by simple and well-tried tests. He records whether certain signs are present or absent. After a time the routine becomes automatic. Among trained physicians the good and the not-so-good both apply it.

The difference between them is not, or at least not solely, that one has a better examination technique. The chief difference is that one can reason better than the other: he can swiftly and unerringly apply hypothesis and experiment to the facts elicited as his examination proceeds. Clinical examination is never passive but always guided by an inquiring mind. The data first obtained suggest to an experienced clinician a disorder that could account for them: from it he infers that some other signs will, or will not, be present, and he promptly looks for them. If their absence or presence falsifies the supposition he makes another and submits this to the test of further observation. Sooner or later a hypothesis is made which no new observations falsify. A diagnosis is nearly always reached by this process of elimination.

Even though his reasoning may be largely unconscious, the man who detects signs that elude others is usually deliberately looking for them, while they are passively observing. To excel, a clinician must not only have acquired a perfect technique of clinical examination, he must have the ability to reason from his observations, which only wide knowledge and experience of disease, and a retentive memory, can give him.

The technical skill needed for exact observation is the third ingredient in the composition of a good observer; it is here that the naturally observant person has an initial advantage.

The training of an accurate observer consists in learning and consciously practising a routine until it becomes an unconscious habit. The routine consists of two parts:—

1. Learning the details to be observed of the particular object studied, and observing them in a definite order which is never varied. Only by *systematic observation* will nothing of importance be overlooked. A radiologist does not look at radiographs of the skull with a blank mind, waiting receptively for some abnormality to strike his consciousness, in no matter what order. He looks inquiringly at the films in his own predetermined order: say, first, the structural pattern of the bone of the vault of the skull, then perhaps the contour of the bones, the sinuses, the sella turcica, and so on. The parts and the details and the order in which they are studied naturally vary with the object studied, and probably with the observer, but a systematic routine there must be.

2. Learning to *record what is observed*, either by making notes or sketches. So can be avoided the besetting sin of all untrained people, of failing to distinguish what they observe from what they infer from their

observations. It may be reasonable for a trained bacteriologist to look through a microscope and say he sees the diphtheria bacillus, but it would not be reasonable for a learner. All that either of them sees is a bacillus of certain morphological and staining characters. It is never right for a radiologist to say that tuberculous lesions can be seen in a chest film, or to assert from a study of his barium-enema films that there is a new growth in the colon.

It is particularly important for a student of a biological science to cultivate the habit of making detailed notes of his observations, and whenever appropriate of making simple line drawings. Nothing else will so surely release him from the universal tendency to confuse facts observed with inferences from them.

Errors of Observation

Everyone makes mistakes, not only in reasoning but in the observations on which our judgements are based. All observation, whether qualitative or quantitative, is subject to error.

Man's first observations of the world around him were qualitative: he noticed the existence of things and their qualities, of events and their regularities and associations. Mistakes in this kind of observation are likely to be made by an untrained observer, whose sense impressions are often imprecise. For we must remember that every observation depends for its accuracy on the proper functioning of one or more of the senses, or of some instrumental aid to perception. A beginner in the study of medicine, for instance, listening to a heart, is often unable to detect an abnormal heart-sound. He may say that there are two sounds, when an experienced observer will say that there are three; and he himself may come to admit, after some practice, that there are in fact three. The same applies to sight and touch. The student may not see cyanosed lips or clubbed fingers; or he may not feel the edge of the liver through the abdominal wall. But with practice sensation becomes sharper and more discriminating.

Beginners need not be unduly discouraged. The complacency of experienced observers has been badly shaken in recent years, for it has been shown that inconsistency among them, however highly trained and experienced they may be, is quite common. Even the same man is liable to be inconsistent in his observations of the same thing on different occasions.

Careful tests have shown that even an experienced radiologist, asked to pick out cases of pulmonary tuberculosis from a large number of mostly normal chest films, will overlook about a quarter. Those that were missed were not all border-line cases. It is recorded that, given a second chance, the observer immediately detected the abnormality that he had overlooked.

Such inconsistencies can, of course, be due to difficulty in interpretation of what is seen; what is a very slight abnormality may be judged normal. But, in most of the instances of this test, abnormal shadows were seen on one occasion and not on another; it was wholly a matter of faulty observation. An X-ray film of the chest records the shadows thrown by the heart and its great vessels, the air passages, the structures at the root of the lungs, the bronchial tubes and blood-vessels caught at every angle by the X-rays that pass through the chest to the film, and the alveolar structure of the lung itself. To a beginner it is a very confusing collection of shadows; to an expert it is usually apparent at once whether a shadow is normal or not. In this study experts repeatedly missed seeing what was at once recognized to be an abnormality on a second occasion.*

To look at an X-ray film and see a glaring abnormality when previously you had seen nothing but normal appearances is a bewildering and rather alarming experience. You cannot account for the lapse, unless it can be ascribed to fatigue.

Inconsistency among experienced medical observers is not confined to radiologists. In another inquiry into 'observer error' some experienced physicians were asked to examine a number of patients and to say whether certain 'physical signs' were present or absent. They were not asked to interpret what they observed; they were simply asked to say whether abnormal redness of the throat, or cyanosis of the lips, and so on, were present or not. In every instance inconsistencies were revealed.

This variability among experienced observers has attracted especial attention in the last decade or so because of the prominence in research of group experiments, in which rigid criteria are needed, both for the diagnosis of the disease and for signs of improvement or deterioration. If diagnosis for a research worker is to be made on a single test or on a very few, or if there is only one criterion, say, an X-ray appearance, for a pathological change, or for disappearance of such a change, then observer error becomes so important that it may vitiate the whole inquiry.

All research workers in medicine are now alive to the possibility of errors of observation. If the study will be profoundly affected by such observations as the recognition of X-ray changes, the presence or absence of physical signs in clinical examination, or in the recording of some instrumental test, the remedy is to have the test done by two or more independent observers.

In other sciences, and increasingly in the biological sciences, fundamental observations are quantitative. But these no less are subject to error.

Observations of degree must soon have followed man's first notice of the existence of things. One thing was noted to be heavier than another, or longer, wider, harder, louder, hotter, and so on, and roughly by how much. These were the first crude measurements, for all measurements record a ratio. Gradually, by the perfection of measuring instruments, they were made more and more precise. Precision has been brought to crude quantitative comparisons by the use of numbers, either for counting or for recording the ratio of a magnitude to a fixed unit.

---

* A full bibliography on observer error is given in WITTS, J. J. (1959), *Medical Surveys and Clinical Trials*. London: Oxford University Press.

The basis of all measurement is the observer's ability to recognize that two things are equal. What we judge to be equal are the magnitude to be measured and a subdivision or multiple of a standard unit. We match a length to be measured with a subdivision or multiple of a standard unit of length, and recognize that the two are equal by a coincidence of two marks. In weighing, we match the body to be weighed with a subdivision or multiple of a standard unit of weight, deciding that they are equal when two marks on the balance coincide. Colorimetric estimations in biochemistry ultimately depend on the observer's judgement that two colours are equal in tone.

A measuring instrument may be constructed to compare the magnitude to be measured with a standard unit by first converting both into linear magnitudes. Blood-pressure is measured by measuring the length of a column of mercury; time intervals may be recorded by a marker on a moving strip of paper, and measured as lengths; small electric currents can be compared by comparing the movements they can be made to impart to a needle, first converting these into corresponding linear magnitudes.

An *error of measurement* is a deviation from the true value. We will leave for the moment the meaning of 'true valuc'. The cause of such deviations is in the observer, or in his measuring instrument, or in the technique of the particular measurement, or in the environment.

Let us think of successive measurements of the same thing, by the same observer, with the same instrument. A supernatural onlooker who knows the true value might well see that every measurement is being made too low, or too high. Some men are prone to this kind of error. Some measuring instruments also may have the same defect, constantly causing an error in one direction—too low or too high.

These errors are usually referred to as *constant errors*: errors that are constant in that they always lie in one direction—above or below the true value. They have the same magnitude and sign in all observations.

It is not difficult to see a number of possible reasons why any given measuring instrument, or technique of estimation, should be subject to a constant error. But the observer? It may, of course, be ascribed to lack of skill in the particular technique. It may be due to bias, if the observer knows what results he wants. Unconsciously, or so we say, we all tend to make readings favourable to our own particular views. In research the only safeguard is not to know what results are favourable.

If the observer has checked his instruments, eliminated personal bias, and perfected his technique, there remain the *random errors* that no one can avoid. Successive measurements of the same thing will not be all the same. Even if it is the same observer who carries out repeated measurements, he will not get precisely the same results each time, though most of his readings will be close together.

So much, of course, is well known to all investigators: everyone tries to eliminate these errors by taking the mean of several discrepant readings. A 'random error' is a deviation of a measurement from the mean.

An observer may be compared to a marksman: where the bullets hit the target, to his observations. The centre of the target is the true value. The distance from the centre of a place where the bullet hits is a random error. The hits will be grouped around the centre much as shown in the diagram:—

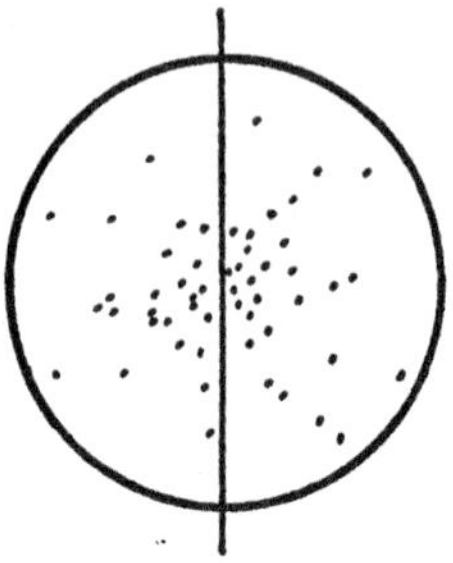

We call random errors that give a reading larger than the mean *positive*, those smaller, *negative*. Or you can call those on the right of a vertical line through the centre of the target 'positive errors', and those on the left 'negative errors'.

It is found by experience that random errors exhibit a remarkable regularity which becomes more pronounced the greater the number of measurements. It is found that:—

1. A positive error and an equal negative error occur with the same frequency.
2. Small errors are more frequent than large ones.
3. Very large positive and negative errors do not occur.

Suppose, to give an imaginary example, we have weighed the same object 500 times: we may find 100 positive and 100 negative deviations from the mean between 0·0 g. and 0·01 g.; 80 positive and 80 negative between 0·01 and 0·02 g.; 40 each between 0·02 and 0·03 g.; 20 each between 0·03 and 0·04 g.; and 10 each between 0·04 and 0·05 g.

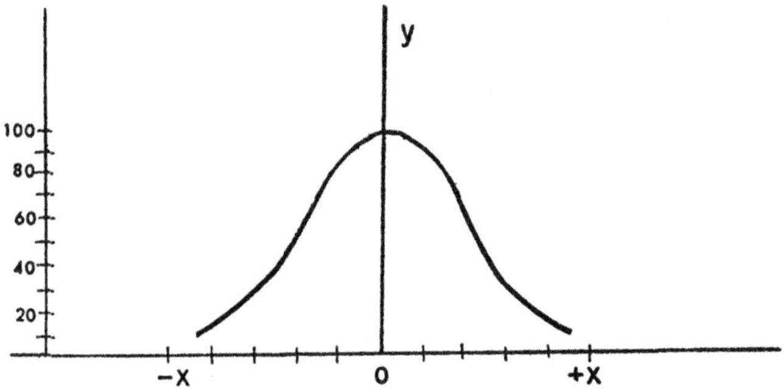

Fig. 1.—Frequency distribution curve for random errors.

We can see the distribution more easily if we plot these values as a curve. Plot the frequency of the errors as ordinates, and their magnitude on the horizontal axis—positive to the right and negative to the left. The curve obtained is shaped like a bell (*Fig.* 1). Strictly it is a polygon, but it approximates to a curve as the ranges of the deviations are made smaller and the number of measurements larger.

If this relation between the magnitude of a random error and the frequency of its occurrence (or its probability) is precise, it can be expressed mathematically: y is a function of x, $y=f(x)$, and this function when plotted must be a curve that is symmetrical (for positive and negative errors occur in the long run with the same frequency); y must be maximum when there is no error; the curve must cut the axis of y approximately at right-angles (for there are a large number of values approximately equal to the mean); and it must slope away at first gradually then rapidly (for as x increases y decreases).

The equation that represents this curve is called the *law of errors*, and the curve the *curve of errors*, the *Gaussian curve*, or the *normal distribution curve*.

At first sight the beautiful regularity of random errors seems unaccountable, but it can be explained by assuming that there is an indefinite and very large number of minute and independent sources of error, some producing positive and others negative errors, and both equally probable. The selection of the mean as the true value rests on this assumption: that unknown very small sources of error will in the long run be as often positive as negative.

The law of errors is a mathematical model which describes with a fair approximation what are believed to be the facts about random errors encountered in the real world. It was deduced, by Gauss and others in the eighteenth century, from certain assumptions, the chief of which was that there is an infinite number of infinitely small causes of error in any act of observation. It is therefore a mathematical statement of an empirical law, and not a theorem deduced from assumptions that we are bound to accept. It has been remarked that everybody firmly believes in the law of errors 'because mathematicians imagine that it is a fact of observation, and observers that it is a theorem of mathematics'.*

Nevertheless, the curve of errors does express with close approximation the relation between the magnitude of a random error and its frequency. When actual errors of observation have been compared with what is required by the mathematical law, agreement has been very close. Jevons gave as an instance of this correspondence 470 astronomical observations of the same phenomenon by Bradley, Astronomer Royal in the early eighteenth century, shown in *Table I*, which is reproduced from *The Principles of Science*.†

An important difference between experience and the mathematical law of errors concerns very large errors. Though, in the curve of errors, when x becomes very large y becomes vanishingly small, the curve never

* POINCARÉ, H. (1892), *Thermodynamique*, quoted by MELLOR, J. W. (1939), in *Higher Mathematics for Students of Chemistry and Physics*. London: Longmans, Green.

† JEVONS, W. STANLEY (1887), *The Principles of Science*. London: Macmillan.

meets the axis of x. This means that theoretically errors however large are possible, though exceedingly infrequent. In practice some very large errors are impossible: if you are measuring the diameter of a penny it is not possible to make an error of 20 miles.

Table I.—A RECORD OF 470 ASTRONOMICAL OBSERVATIONS SHOWING THE RELATION BETWEEN THE MAGNITUDE OF A RANDOM ERROR AND ITS FREQUENCY (*after Jevons*)

MAGNITUDE OF ERROR IN PARTS OF A SECOND OF ARC, BETWEEN:—	NUMBER OF ERRORS OF EACH MAGNITUDE	
	Observed	*Theory*
0 and 0·1	94	95
0·1 and 0·2	88	89
0·2 and 0·3	78	78
0·3 and 0·4	58	64
0·4 and 0·5	51	50
0·5 and 0·6	36	36
0·6 and 0·7	26	24
0·7 and 0·8	14	15
0·8 and 0·9	10	9
0·9 and 1·0	7	5
Above 1·0	8	5

The following property of the normal curve will be referred to later in the book. It is possible to calculate the probability of any error being greater, or less, than any stated amount.

Three definitions must be given:—

1. The *mean* of a collection of measurements refers to the arithmetic mean: it is the sum of all the values divided by their number. If x represents any value and n their number, the mean $\bar{x}$ (x-bar) is written symbolically S $(x)/n$.

2. The *variance* is the mean of the squared errors (or squared deviations): S $(x-\bar{x})^2/n$.

3. The *standard deviation* is the square root of the variance:

$$\sqrt{\{S\ (x-\bar{x})^2/n\}}.$$

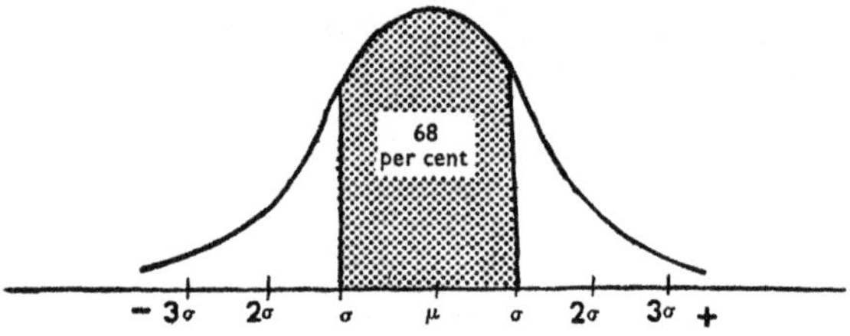

Fig. 2.—The normal curve.

In *Fig.* 2 a curve of errors is shown, with multiples of the standard deviation, denoted by σ, from the mean of the values μ, on the horizontal axis—positive to one side, negative to the other.

The area enclosed by the curve and the horizontal axis represents the total number of observations, and the fraction of this area standing on any part of the horizontal axis, say, μ to σ, represents the fraction of the total observations that occur between these limits.

It can be shown that (approximately):—

68 per cent of the observations are within the range $\mu \pm \sigma$;
95 *per cent of the observations are within the range* $\mu \pm 2\sigma$;
99 per cent of the observations are within the range $\mu \pm 2{\cdot}6\sigma$;
99·7 per cent of the observations are within the range $\mu \pm 3\sigma$.

It will be seen that the probability of a measurement being outside the range $\mu \pm 2\sigma$ is 0·05, and that the probability of its being outside the range $\mu \pm 2{\cdot}6\,\sigma$ is 0·01.

CHAPTER XV

THE SEARCH FOR CAUSES

IT is not possible, most scientists believe, to teach a man how to do research. You may suggest a few guiding principles: he is then on his own. Writings on the logic of scientific method, they say, have been of little or no help to anyone in the work of discovery.

Perhaps one helpful thing that logicians have done is to analyse the procedures and the reasoning scientists have adopted in their inquiries into the causes of things. Francis Bacon began it. His ideas were developed by Hume in the eighteenth century, by J. S. Mill last century, and by many successors in more recent times. Naturally they begin when the scientist has finished; consequently their analysis has made research into the causes of things seem a much simpler business than it really is.

Knowledge of what is evidence for two things being causally related, and what is not, is less common than it should be. If proof of this is needed, look at the difficulty some people profess to have in deciding whether smoking cigarettes and lung cancer are causally related or not. What I wish to suggest is that, to know if our conclusions are valid, and what gives them validity, must be a help in all inquiries into causation. And I wish to discuss the most useful and commonly applied criterion.

The problem I wish to consider is not that of diagnosis: the discovery of the cause of an illness, a death, a fire, an aeroplane crash, or what. This problem simply involves the elimination of one known possible cause after another, until one is found which cannot be excluded. My purpose is rather to examine how such causes have become known, or, more generally, to examine methods used by scientists in their efforts to reveal the cause of a class of things of completely unknown origin, and to determine when it can be said that evidence of causal relation has been found.

We may take as an example an inquiry into the cause of a disease of unknown origin. The story of how the cause of yellow fever was discovered will serve as well as any.

For more than 200 years yellow fever had been one of the great plagues of mankind. Repeatedly sweeping in appalling epidemics over the West Indies, parts of tropical South America, and the West Coast of Africa, 'yellow jack' was dreaded by sailors on their way to stricken ports. In 1900 an epidemic was raging in Havana, where units of the American Army were stationed.

The infection struck suddenly, with high fever. About 1 in 4 had a short fulminating illness: racked with pain, vomiting black blood, sinking with deepening jaundice, they were dead in a week.

The American Army appointed a research team to investigate the problem: Major Walter Reed, Dr. James Carroll, Dr. Jesse Lazear, and Dr. Aristides Agramonte. Almost immediately Reed and his associates adopted a hypothesis first suggested in 1881 by a Cuban doctor, Carlos

Finlay, that the disease is transmitted by a mosquito. Finlay had noted that the only blood-sucking insect common where epidemics occur was a distinctive species of mosquito, easily recognized by its peculiar thoracic markings, now called *Aedes aegypti.*

In their first series of experiments they allowed mosquitoes to bite patients stricken with the illness, then, after a variable time, to bite volunteers—men who had never had the disease—for even a mild attack confers lifelong immunity.

They did not try this on one man only. They knew that there were variables beyond their control that would make a negative result to one experiment valueless. They knew, for instance, that a volunteer who had lived in a yellow-fever district might well be immune without knowing it. They therefore observed the effect of 'contaminated' *Aedes* mosquito bites in one volunteer after another. In the first nine it was without effect. The tenth was Carroll. He became severely ill with the fever, dying from its effects several years later. The next was Lazear, another of these incomparably courageous investigators: after a short fulminating illness he was dead. Still men volunteered—as many as were needed.

This preliminary experiment clearly justified a more elaborately planned study of the same kind. 'Camp Lazear' was built near Havana, with mosquito-proof houses. Volunteers were selected from the American army of occupation. When one left the camp he was not allowed to return; another non-immune volunteer replaced him.

In the first 3 months 10 volunteers contracted yellow fever from the experimental bite of an infected *Aedes* mosquito. As soon as a man became ill he was taken to a yellow-fever hospital to prevent the disease spreading in the camp.

Two more variables affecting the outcome were soon recognized. For successful transmission of the disease it was necessary, first, that the insect should have sucked the blood of a yellow-fever patient during the first 4 or 5 days of the illness; after that it was too late: secondly, it was necessary that an interval of at least 12 days should elapse before the infected mosquito was allowed to bite a volunteer—clearly the minimum time for the virus to develop sufficiently in the insect.

When these requirements were adopted failure was the exception.

Another experiment carried out at Camp Lazear was designed to falsify the alternative theory, widely believed by the native population, that yellow fever was contagious and was transmitted by direct contact with patients or with their clothing or bedding. Three non-immune volunteers lived for 20 days in a mosquito-proof house where the soiled clothing, bedding, and other personal belongings of yellow-fever patients had been collected. Each day these men thoroughly shook out the clothing and bedding, and scattered them about their rooms; then they slept in soiled sheets and blankets. No one contracted the disease.

For General Gorgas, Chief Sanitary Officer of the American Army in Havana, this 'cause' was good enough. He had no need to wait for the full story. The epidemic was still raging in 1901. And so the work of eliminating the *Aedes* mosquito was started. Sanitary inspectors policed the district to see that stagnant water was not allowed to collect near dwelling-houses; rain-water butts were protected; on suspicion of yellow fever a patient was screened from mosquitoes. In the next 7 months only 11 deaths from yellow fever occurred; in no similar period in the previous 150 years had there been less than 100. From September, 1901, to July, 1902, not a single case of yellow fever occurred in Havana. It is now a rare disease everywhere.

There was, of course, much more to be done. What was it that the mosquito transmitted? It was clearly an infective agent contained in a patient's blood during the first 4 or 5 days of his illness. One of Carroll's experiments was this:—

Blood from a yellow-fever patient, though containing no bacteria that could be seen or grown artificially, had given the disease to healthy volunteers after inoculation. At this time the existence of living things even more minute than the smallest bacteria had just been discovered—the *filterable viruses*. These ultra-microscopic organisms, which would pass through porcelain filters fine enough to hold back the smallest visible bacteria, had been shown to be the infective agent in two diseases: tobacco mosaic, and foot-and-mouth disease of cattle. With this in mind, Carroll filtered the yellow-fever blood through Chamberland's filter and injected the clear filtrate, presumably free from any bacteria that might have been overlooked, into 3 volunteers. Two of them developed the disease; the clear filtrate was infectious like the unfiltered blood. Yellow fever was the first instance of a human disease proved to be caused by a virus.

It looked as though knowledge of the cause of yellow fever was complete. But it was not. The full story was not told until about 1940. Here I shall give merely some salient features of the work that followed the beginning made by Reed and his colleagues, to show the kind of laborious task that research into the causation of a disease may be.

In 1918 epidemics of yellow fever were still occurring in Central and South America and in Africa. The Rockefeller Foundation decided that its eradication throughout the world was practicable and should be undertaken.

While working at the Yellow Fever Hospital in Guayaquil, as one of the Foundation's team which eradicated the infection from Ecuador as successfully as it had been from Havana, Noguchi, one of the world's greatest bacteriologists, reported his evidence that a *spirochaete* was the cause of the disease. Years later it was clear that some of the patients he had investigated had suffered from Weil's disease, an epidemic jaundice clinically almost indistinguishable from yellow fever.

Modern research on the problem began in 1925 when the Rockefeller Foundation sent a commission to the west coast of Africa. Noguchi's error had persisted, as the errors of eminent authorities frequently do, and everyone now believed the cause of yellow fever to be his *Leptospira icteroides*.

The commission's first task was to decide the issue, and this would have been quite impossible, as would all their future work, if they had not found a laboratory animal susceptible to the disease. Reed's experiments had inevitably been restricted; he had to use human volunteers as sparingly as he could.

In 1927 the infective agent was successfully transmitted to a rhesus monkey. It could easily be passed from monkey to monkey by inoculation of blood taken early in the disease, or by the agency of the *Aedes* mosquito.

The *Leptospira* could not be detected in any case of African yellow fever. On the other hand, inoculation of monkeys soon confirmed Reed's original discovery that the infective agent is filterable. And it was eventually proved that the yellow-fever virus in Africa was identical with that in South America. But rhesus monkeys are expensive and difficult animals to work with, and in 1931 it was discovered that white mice are susceptible if inoculated intracerebrally. It was now possible for the first time to carry out large-scale immunity tests. After recovery from yellow fever, even though the attack was mild, a man is immune for the rest of his life, because specific antibodies to the virus have been formed in his blood as a result of the infection. If these specific antibodies can be demonstrated in a person's blood it is proof that at some time or other he has had yellow fever.

Mice could be protected against intracerebral inoculation with yellow-fever virus by simultaneous injection of serum from a human being recovered from the disease. This *mouse protection test* became one of the principal means by which yellow-fever research advanced.

In 1932 there was an epidemic of yellow fever among woodworkers in the forests of Brazil, away from human habitation, and where the *Aedes* mosquito is not found. Studies were therefore concentrated on: (1) The presence of the virus in jungle mosquitoes; (2) The presence of specific antibodies in jungle animals, using the mouse protection test.

It was soon found that adult jungle monkeys showed a high degree of immunity; they had therefore at some time suffered from yellow fever. And it was found that a species of jungle mosquito living mainly in the tree-tops could transmit the infection.

By 1940 the co-ordinated research organized by the Rockefeller Foundation had shown that the yellow fever virus is a parasite of monkeys in the tropical jungles of Africa and South America. It is picked up from them by jungle mosquitoes which harbour it for the rest of their lives. Men may still be infected by an insect bite in the jungle and, returning to a town, start, with the help of the domestic *Aedes*, a minor epidemic.

The basic structure of very many experiments was given by Mill:—

> It is the very nature of an experiment to introduce into a pre-existing state of circumstances a change perfectly definite. This change may be taking from it a suspected cause, one at a time, and noting if the phenomenon disappears; or introducing a suspected cause, and noting if the phenomenon occurs.

This design of an experiment to determine causation can be seen throughout the researches just described. Reed and his colleagues introduced a suspected cause into a volunteer and noted if yellow fever developed. General Gorgas, in his sanitary measures for Havana, removed a suspected cause, the *Aedes* mosquito, and noted if yellow fever disappeared from the community. The experiment that falsified the popular notion of contagion was designed on this plan. The experiments falsifying the hypothesis that the infective agent was a bacterial micro-organism, and lending support to the supposition that it was a virus, had the same design. It was used in the discovery of a susceptible laboratory animal, of the mouse protection test, and of all the experiments revealing the existence of jungle yellow fever.

The general design of an experiment in this search for causes is not in doubt. Among criteria given by most logical writers for the causal relation are the following:—

1. If, in a series of instances, in a complex situation the phenomenon *B* occurs, and if withdrawal of a suspected cause *A*, *alone*, is always accompanied by the withdrawal of *B*, then it is probable that *A* is causally connected with *B*.

2. If, in a series of instances, in a complex situation the phenomenon *B* does not occur, and if the introduction of a suspected cause *A*, *alone*, is always accompanied by the occurrence of *B*, then it is probable that *A* is causally connected with *B*.

This reasoning is valid but practically useless.

It is useless, first, because its validity depends on introducing or withdrawing *A*, and *nothing more*. There must be no other change brought to the situation and this is extremely difficult or impossible to arrange.

We can never know that the introduction of a suspected cause, *A*, is the only change brought to the situation. Suppose, if you can, that Reed had given a volunteer a currant bun, believing he had done nothing more, and that the man had developed yellow fever. A conclusion of causal relation would at once be seen to be the *post hoc ergo propter hoc* fallacy. The man may have been bitten by a mosquito while he was eating his bun.

We can never be sure that the single change that we think we are bringing to a situation is in fact the only change, especially if the situation is a living thing. Living things are so inherently variable that changes in them that may affect the outcome of the experiment may be going on all the time.

I shall defer, for the next chapter, a discussion of how, to some extent, we can overcome this difficulty of variability, by 'experimental controls'.

The second reason why this criterion is useless is that we hardly ever get such uniformly positive results in our experiments. It is not true that a 'cause' cannot be present when the phenomenon is absent, or absent when the phenomenon occurs. It depends on the kind of 'cause' we are talking about.

A is said to be a *sufficient cause* of *B*, if all we are saying is that *A* implies *B*, and *B* follows *A* in time. A bite by an infected *Aedes* mosquito is a sufficient cause of yellow fever in a non-immune person.

A is said to be a *necessary cause* of *B* when what we mean is *B* implies *A*, and the same time relation holds. The *Aedes* bite in the above example is not a necessary cause of yellow fever: the disease can be contracted in the jungle where there are no *Aedes* mosquitoes. Inoculation of the yellow-fever virus is a necessary cause of yellow fever.

The cause envisaged in the Law of Universal Causation is a *reciprocating cause*: *A* implies *B*, and *B* implies *A*. It is the earlier stages of the whole process that constitute the phenomenon and we can go back to its antecedents as far as serves any useful purpose. In this sense the cause of an epidemic of yellow fever, to single out the salient features from a mass of detail, is the existence of a particular virus in jungle monkeys, its spread to some man by a jungle mosquito, the fact that this victim reaches a town within 5 days of his illness, and the fact that there are in this town *Aedes* mosquitoes which transmit his disease to others who lack immunity to the virus.

A mistake we make is to forget that most writers on logic are talking of causes of these kinds, and we, as a rule, are not. A mistake some logicians have made is to imagine that any scientist would begin his investigation by looking for such a cause.

A research worker engaged on a problem of causation first resolves the problem into its simpler parts. The 'cause' of logic is a constellation of popular 'causes', which his studies will be designed to reveal, one by one. Most of these are what we call 'causal factors'.

It is now common practice to speak of a condition that we believe to stand in causal relation to a phenomenon, but which is clearly neither a sufficient nor a necessary cause of it, as a 'causal factor'. By this we mean that it is one of a group of conditions that collectively is either a sufficient or a necessary cause, or both; or, vaguely, that it may contain, or be closely associated with, an unknown x that is one of such a group. A stagnant pool in Havana in the year 1900 was a causal factor in the epidemic. Cigarette smoking would now be admitted by most people to be causally connected with the production of lung cancer. It is obviously neither a necessary nor a sufficient cause: we therefore call it a 'causal factor'.

In an inquiry into the cause of a disease, be if of human beings, or farm animals, or growing crops, we may introduce a suspected causal factor into a number of subjects. The probability that, if it is a causal factor, the disease will always follow is so infinitesimal that we may say it never happens.

The useful criterion for causal relation is this: If, in a series of subjects, we introduce a suspected causal factor, A, and find that the disease, B, appears significantly more frequently than in a similar group of untreated subjects, then A is probably causally connected with B.

Statisticians have shown how we can decide what difference is significant, and what is not. The principles of statistics have become better known to medical and biological research workers during the past 25 years or more; and so increasing use is being made of these group studies in research (*see also* Chapter XVII).

Experiments, in the ordinary sense of the word, are not always necessary. We may simply collect instances of the phenomenon and look at the antecedents to see if a suspected cause was present with greater frequency than in a comparable group in which the phenomenon is absent. Alternatively, we may study two groups of subjects exactly alike in all relevant respects, except that in one group the suspected cause is present, in the other absent; and we may then wait to compare the frequency with which the phenomenon follows in the two groups. The two methods of inquiry are called respectively the *retrospective* and the *prospective methods*.

In the *retrospective method* we start, for example, with instances of a disease whose cause we are seeking. We find by inquiry the frequency of exposure to a suspected cause. Suppose that, in a study group of people affected by a disease *a*, a suspected cause *A* has been present among the antecedents in 50 per cent. *A* could be a causal factor in the production of *a*. But this would derive no credibility from the observation if, in a control group of people not suffering from the disease, *A* was present among the antecedents with the same frequency. Suppose, however, that in the control group the frequency of past exposure to *A* was found to be 30 per cent, or 20 per cent, or 10 per cent: does there come a point when we can say that the difference is too big to have occurred by chance?

Any such difference between the two groups could, of course, be a chance difference. Even if the suspected cause is innocent, having no influence whatever on the production of the disease, the frequency of its occurrence in the two groups may still differ widely, and this purely by chance. But it is always possible to calculate the mathematical probability of any given difference occurring by chance. If the difference is a very big one, the probability of its having occurred by chance will be very small. If the two frequencies differ only slightly, the probability that it is a chance difference will be very big.

What the research worker does is to accept a convention. If the probability that a difference he has found is due to chance is below an accepted level—if, say, the chances against it are 99 : 1—he will reject this interpretation. He will accept the alternative, that there is a high degree of probability that the difference is due to something other than chance, and proceed to consider what this may be.

The *prospective method* of inquiry proceeds in the other direction. We begin by noting occurrences of a suspected cause, *A*; subsequent observations show the frequency with which the expected effect *a* follows. A control group is composed of people in whom the suspected cause has not appeared, and the same follow-up is organized, to find the frequency with which the same phenomenon *a* occurs later.

If the difference between the two groups in this respect is found to be so big as to be unlikely to be a chance difference, we have the same problem of interpretation.

The advantages lie with the prospective method, but both are commonly employed in the same inquiry. As one of the best illustrations of their combined use we may take the inquiries of Doll and Bradford Hill into the association of smoking and lung cancer.

In their first series of observations, which were retrospective, 20 London hospitals were asked to report all patients admitted with carcinoma of the lung. Each patient was then paired with a control patient, suffering from any disease that was not cancer, of the same sex and same 5-year age-group.

The smoking habits in every case were recorded, so that each member of both study and control group could be classified into smokers and non-smokers; and the smokers into various subgroups, including groups smoking an average of increasing numbers of cigarettes daily—less than

5, 5–14, 15–24, and so on. In their report of 1952 the smoking habits, judged by various measures, of 1357 men with lung cancer were compared with those of a control group of 1357 men with other diseases. Differences such as these were found: 7 of the lung-cancer group were non-smokers, 61 of the control group; 293 of the lung-cancer group smoked 25–49 cigarettes daily, 154 of the control group; 38 of the lung-cancer group smoked more than 50 cigarettes daily, 12 of the control group. Every comparison made, by whatever measure the amount of smoking was estimated, showed a disproportion of the same significance.

The probability that the difference found was due to errors of sampling, smoking habits being entirely unassociated with lung cancer, was often as low or less than 0·001, i.e., it was a difference that, in every thousand such pairs of samples, would occur, in the long run, once.

In addition to these earlier studies, the prospective method of inquiry was used. Questionnaires were sent to all doctors in the British Isles, so that they could be classified into smokers, ex-smokers, and always non-smokers; the smokers and ex-smokers were also classified into various subgroups according to the amount and type of smoking. The inquiry continued by collecting information about deaths. Because of a ready availability of records about doctors, this was easy, and at the end of 6 years it was felt that scarcely any had been missed.

From these data it was possible to calculate the death-rate from lung cancer in non-smokers and in the several categories of smokers. The following show the kind of differences found: among non-smokers the death-rate from lung cancer was 0·07; among cigarette smokers averaging 1–14 daily it was 0·95; averaging 15–24 daily it was 1·67; averaging 25 or more daily, 2·76.

Again, the differences that were recorded could be due to chance with a probability often of 0·01, at times as low as or lower than 0·001.

In both methods a disproportion of the frequency in the study group compared with the control group, too big to be reasonably due to chance, must be due to one of two things: (1) A control group that is not exactly comparable; (2) An association of the phenomenon and its suspected cause through some fact of causation.

It is by no means easy to select a control group that is exactly comparable to the study group in all those characters that can affect the outcome of the study. A difference in the composition of the control group other than the absence of the phenomenon under investigation, or the suspected cause, may affect the comparative frequency that we propose to determine. Each study-group subject should therefore be paired with a control-group subject as regards sex, age-group, and any other character thought to be relevant.

There are many pitfalls. If they are not avoided, the end of a long and arduous study may bring no more than entirely artificial differences, incapable of useful interpretation. Planning these experiments is not easy.

If we can accept that a difference in frequency in study and control groups is not due to chance, and if we have taken pains to make the groups strictly comparable, then there is only one interpretation—the phenomenon and the suspected cause are in fact causally associated.

It will be well, however, to review what can be meant by this. In the lung-cancer studies the fact that many cigarette smokers (*A*) do not die of lung cancer (*a*), and that lung cancer occurs in non-smokers, completely eliminates the possibilities:—

1. *A* is a sufficient cause of *a*.
2. *A* is a necessary cause of *a*.
3. *A fortiori*, *A* is a sufficient and necessary cause of *a*.

We are left with:—

4. *A* is one of a group of causal factors that is a sufficient cause of *a*.
5. *A* at times contains an unknown *x*, that causes *a*, or is one of a group of causal factors that causes *a*, either as a sufficient or necessary cause or both. For example, *x* may be an unknown carcinogen in tobacco smoke, inhalable in other ways.

In this kind of observation, where the suspected cause occurs naturally in a group of subjects and is not artificially put there by us, there is one further possibility:—

6. An unknown may cause both *A* and *a*. In the lung cancer problem this could be, however improbably, the stress of life causing an immediate tendency to smoke cigarettes, and a later one to develop lung cancer.

CHAPTER XVI

CONTROLLED EXPERIMENTS

An experiment is an observation planned to answer a question—a question usually asked because it is a consequence of some theory we are trying to verify.

If the answer is what the theory requires it is to be interpreted as evidence in favour of the theory. No such interpretation is possible, however, if there is something in the experimental procedure that could account for the result just as well. A tiro, unless shown how to do it, might find it difficult to provide evidence in favour of an assertion that you can tell margarine from butter. For whatever successes he recorded, a critic could say that they were due to chance.

In every biological experiment some alternative explanation of the results lurks in the experimental procedure. The experiment must be planned in such a way that this can be shown to be, if it is so, improbable or virtually impossible, otherwise the results will be worthless and the experiment might as well never have been performed.

Any device which achieves this is what is meant by an experimental control.

Take this simple example. Will the yellow-fever virus living in an infected mosquito pass through a bacterial filter? This was one of the questions asked by the West Africa yellow-fever team, once they had discovered that the rhesus monkey was susceptible to the virus and therefore provided a test for its presence in any fluid.

The virus in the blood of infected monkeys would pass easily through various grades of bacteria-proof filter, being stopped only by a very fine-grade collodion membrane. But when they came to test mosquitoes known to harbour the virus, by grinding them up in normal saline and passing the suspension through the same filters, the remarkable observation was made that the virus was apparently stopped by all of them: no test monkey was affected by any filtrate.

For the moment they had forgotten the critic who would say: 'How do you know that the virus was not killed when the mosquitoes were ground up in saline?' The necessary control was, of course, soon applied, but a less experienced investigator might not have spotted the omission so quickly. Who would suspect that normal salt solution would be a deadly poison to any living thing?

The controls required, which should have been set up with the original experiment, are, first, when testing the filtrate for the presence of active virus to test also the unfiltered suspension, and, second, to test suspensions in a series of different diluents. It was in this way soon established that this is what had happened: the mosquito virus had been killed by the saline in which it had been suspended. When suspended in serum the virus was not killed and passed readily through the bacterial filters exactly as did the virus in an infected monkey's blood.

To an experimenter no criticism is so devastating as the accusation that his *controls* are inadequate. Nothing so discourages the beginner or so provokes the experienced: it implies that they do not know their job. There are men whose chief delight is to set others back by harsh criticism, and the reader may like to know how, in this point of adequate controls, they may be thwarted.

In the experimental plan described by Mill we are to introduce into a situation 'a perfectly definite change', and note the result. This is probably a fair account of what was done to establish the simpler laws of physics: it is an oversimplified statement of how work in any biological research laboratory is planned. There are three main hazards, three sources of an alternative hypothesis that could account for the results, and which therefore impose considerable modifications on this basic plan.

First, there is the difficulty of introducing 'a perfectly definite change' and doing nothing else that could affect the result.

Secondly, Mill's 'situation' is usually a living thing, subject to unknown and unavoidable changes in its *milieu intérieur* that could affect the outcome of the experiment.

Thirdly, the experimental results may represent no more than chance differences between the small samples with which we are compelled to work. If two angels guide the path of man, one goddess makes their task more toilsome if more interesting—Fortuna, *dea varia, lubrica, et fragilis*; the experimenter must know that she is always about, sitting on her wheel, as ready to lure him off course as to lend a hand.

Usually it needs great ingenuity, if it can be done at all, to bring only one change to any situation. Pasteur achieved it in some of his experiments.

His alternative to the hypothesis of spontaneous generation was that, when growth occurs in a fluid culture medium, the micro-organisms have grown from others like them that have gained access with the dust of the air. It was a valid prediction that if you exclude the dust of the air from a culture medium that initially contains no micro-organisms, growth will not occur.

An experiment purporting to show this had been known since Spallanzani first performed it a century before Pasteur. A little fluid culture medium is put in each of two flasks and boiled. One flask is left exposed to the air: the fluid quickly becomes cloudy from a growth of micro-organisms. The neck of the other flask is sealed immediately after boiling, and so the dust of the air is excluded: growth never occurs.

The exponents of spontaneous generation pointed out the weakness of this experiment—another hypothesis that could account for the result and nullify the significance it was supposed to have. You have not only excluded the dust, they said, you have excluded oxygen; and this is needed for spontaneous generation.

Pasteur countered this by an elegantly simple modification of the experiment. Instead of sealing the neck of one flask, he drew it out to an S-shape (*Fig.* 3). Air returned to both flasks during cooling, but in the S-flask the dust was trapped in the dependent part of the swan-neck.

Growth never occurred under these conditions, no matter for how long such flasks were left.

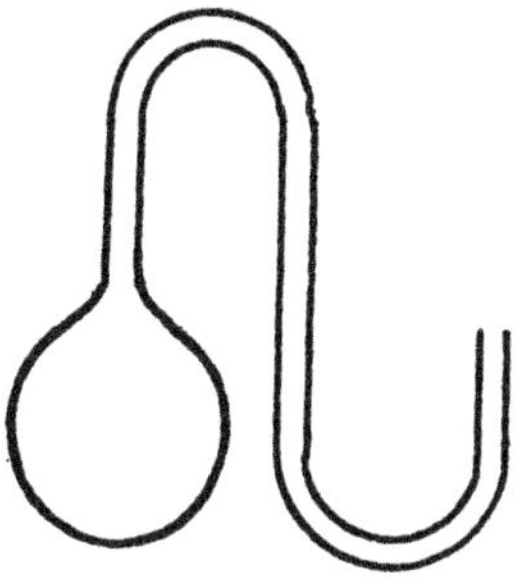

Fig. 3.—The shape of the flask used in Pasteur's experiment.

Into this stable situation Pasteur then introduced the dust of the air, and nothing else: it could be done quite simply by shaking the flask so that some clear fluid spilled over into the swan-neck. In every case growth occurred very soon after this.

Usually it is not possible to introduce a change, *A*, without at the same time introducing, or risking the introduction of, other changes, *B*, *C*, . . . Particularly when the subject of the experiment is some living thing may it be affected by unknown factors in the environment, as well as by what you do to it. It is often possible, however, to introduce *B*, *C*, ... into another subject, and show that they are without effect.

The first experiment that extended our knowledge of tuberculosis illustrates this. Early in the nineteenth century the possibility that contagious diseases were caused by micro-organisms was widely known. About 1840 Robert Koch suggested that one of the criteria for regarding a micro-organism as the cause of a disease should be that it must, if injected into a healthy animal, reproduce the disease. Jean-Antoine Villemin (1827–92), a professor in a French Army medical school, assumed that tuberculosis, known to be contagious, was caused by an unknown micro-organism contained in tuberculous material. The following is a quotation translated from his *Étude sur la Tuberculose*, published in Paris in 1868:—

> March 6, 1865, we took two rabbits about three weeks old. . . . In one of these rabbits we introduced into a little subcutaneous wound behind each ear two small fragments of tubercle, and a small amount of purulent liquid from a tuberculous cavity, removed from the lung and the intestine of a consumptive who had been dead 23 hours. June 20, that is at the end of 3 months and 14 days, we sacrificed it and noted: . . . tubercular implants along the greater curvature of the stomach. The lungs are full of large tubercular masses. The other rabbit, who had shared the same conditions of life with the inoculated rabbit, was then put to death and did not show a single tubercle.*

In this experiment it will be noted that Villemin did more than inject a rabbit with material suspected of containing the infective agent of

* Quoted from MAJOR, R. A. (1939), *Classic Descriptions of Disease*. Springfield: Thomas.

tuberculosis (*A*). There was the possibility to consider that if the rabbit did develop tuberculosis it could have been because of some noxious agent in the environment (*B*, *C*, ...). To show that this was without effect he observed also another rabbit of the same age and sharing the same conditions of life with the inoculated rabbit, but not inoculated.

The *device of pairing* is the first modification of Mill's simple experimental plan. It is imposed by the need to ensure that no unnoticed addition to what we have done has had any influence on the result.

Our observation must be made on a pair of subjects, one treated, the other not; and they must live as nearly as may be under the same conditions. It is comparatively easy to ensure the same conditions of life for laboratory animals; but the best we can do rarely gives exactly the same conditions to growing plants. Suppose the question is whether a particular fertilizer increases the yield of a farm crop. We shall plan to contrast two crops, from the same batch of seed, one treated by the fertilizer, the other not. But we must grow them on adjacent plots, and adjacent plots may not have soil of precisely the same fertility, nor may the two crops get exactly the same amount of sun.

Errors from this remaining variability will be taken care of by a further elaboration of the experimental plan.

The second hazard to be faced in planning an experiment is the confusion that the variability of the subject itself may bring, if it is some living thing. There are two controls by which an experimenter can ensure that his results are not worthless on this account: both may be possible.

First, he can try to secure the utmost uniformity in the pair. Ideally they should be exactly alike in all respects that could affect the experimental result. Such uniformity is difficult to obtain. In some animal experiments the pair contrasted can be from the same litter. But the best we can do rarely gives complete uniformity between the two members of a pair. It never does when the experiment is on human beings, but so far as possible they are paired according to sex, age, occupation, and anything else that we think might affect the outcome.

Secondly, and this is an equally if not more important device, he can extend the observations to a number of pairs, allocating the 'treatment' in completely random fashion to one or the other member of each pair. In this way the error from variability is made to fall equally on the treated and untreated groups.

Uniformity is only required between the two members of a pair that are to be treated differently and contrasted. It need not extend to the parallel pairs of the experiment, for the control is not in the replication of observations but in the *randomization* of 'treatment'. Pairs need not be of animals from the same litter provided those of any given pair have this uniformity.

Experiments of this design have been carried out so extensively, and for so long, that it seems almost unnecessary to give examples. But for the sake of those who are not very familiar with experimentation, here is a brief account of Pasteur's experiment demonstrating that sheep could be protected from anthrax.

Anthrax has been a menace to cattle and sheep from time immemorial; a century ago it was a scourge of man and beast throughout Europe. Many thousands of animals died from it every year. The animal stops in its tracks, shaking and gasping, and bleeding from the mouth, stricken with a fulminating mortal illness. Within a few hours it is dead.

The bacillus of anthrax was discovered by Robert Koch in 1876. Pasteur at once conceived the idea that animals could be protected by inoculating them with attenuated cultures of the bacillus, exactly as he had already immunized fowls against 'chicken cholera'.

One of the most famous and melodramatic experiments in the history of bacteriology was staged at a farm some 30 miles from Paris in 1881. On 5 May, in the presence of a concourse of senators, deputies, veterinary surgeons, physicians, and others, Pasteur divided sheep placed at his disposal into two groups, of 25 each, and inoculated with anthrax vaccine all the sheep of one group. This was repeated some two weeks later. The sheep of the other group were not inoculated. About a week after the second inoculation all the sheep of both groups were injected with a very virulent culture of the anthrax bacillus.

It will be seen that this conforms roughly to the basic experimental plan just described: two similar groups both at risk from anthrax, one inoculated, the other not.

By 2 June all but 3 of the uninoculated group were dead: the last 3 postponed matters until Pasteur had reassembled his distinguished audience, when they conveniently died in their presence. All the sheep of the inoculated group were in perfect health.

This is the kind of experiment that is continually being performed. The expected result is that a proportion, perhaps all, of the treated group will show the looked-for effect, whereas this occurs in a smaller proportion, perhaps none, of the untreated group.

There are rare occasions when a control group is either unnecessary or impossible, and in its place we can rely, or are forced to rely, on past experience. Pasteur's last and most resounding triumph was his conquest of rabies, and here past experience, thoroughly unsatisfactory though it is, is all we can ever have for comparison with a treated group.

Rabies is the most terrible virus disease known to man. Its one redeeming feature is that it is contracted only from the bite of an infected animal, usually a dog. The virus is in the saliva of the animal, and enters the torn nerve-fibres of the bitten victim. Slowly it spreads up the nerves to infect the brain and spinal cord, and so produces its appalling symptoms. Most of us in England have never seen it, but it is still common in Europe and North Africa, and very common in India and the Far East.

Pasteur succeeded in preparing a vaccine of the attenuated rabies virus from an infected rabbit's spinal cord, with which he could immunize a dog in 15 days. The incubation period in man was known to be 4 weeks or longer, so there was time to immunize bitten people during the incubation period.

It was believed at that time that the proportion of people bitten by a rabid dog who contract rabies was about 40 per cent. There was, and still is, no way of knowing precisely. Now it is thought that the figure was probably nearer 15 per cent. Once Pasteur had found a vaccine which he believed could prevent it, he could not withhold it from anyone, to find out the proportion of untreated people who die: no other means of prevention was known. Now it will never be possible, and we shall never know exactly what the figure is.

By 1935, 57,000 people bitten by dogs had been inoculated against rabies at the Pasteur Institute in Paris; less than 1 in 1000 had died. This means that less than 1 in 1000 had contracted the disease, for once symptoms appear no one recovers. More recent figures are not so good; the mortality is given as about 5 in 1000. But there is another difficulty; it is not known exactly how many of the people inoculated had been bitten by dogs known to have rabies. It seems probable that in only about half of them were the dogs actually rabid. And so the mortality of those really at risk and given anti-rabies treatment is about 1 per cent.

If a mortality of 15 per cent in people bitten by rabid dogs and untreated were known to be fairly accurate, the probability of finding this difference by chance would be infinitesimal. All authorities do believe that anti-rabies treatment is of value, but, as it can never be withheld from anyone bitten by a rabid animal, we can never really know.

The third hazard that awaits the research worker, considering an experiment of this design, is the possibility that chance will explain his results.

Everyone, even untrained in experimental work, has an intuitive and vague idea that the smaller any difference he finds between his treated and untreated groups, the larger the groups should be to make it convincing. The 100 per cent difference found in Pasteur's anthrax experiment leaves no doubt of its significance. But between two groups of only 25 each, a chance difference of 20 per cent, which might in some experiments seem quite important, would not be a very unlikely event. To make a 20 per cent difference convincing the groups would need to be much larger: if they were 100 each chance would be a most unlikely explanation and if 500 each the suggestion would border on the absurd.

It may seem then that the obvious control, to refute a critic who tells us that chance, and not what we did to one group, has caused the difference, is to work always with very large samples. But this could make an investigation exceedingly laborious, if not, for the limited resources of one man, impossible. Usually we are compelled to work with small samples.

We cannot eliminate the intervention of chance. Even so, the goddess must work by rules, and if we understand them it is still possible that the critics may be confounded. 'I think therefore it would be useful, not only to gamesters but to all men in general, to know that chance has rules which can be known, and that through not knowing these rules they make faults every day.'* So wrote one of the early contributors to mathematical probability in the seventeenth century. But this is to anticipate.

---

* Pierre Rémond de Montmort (1678–1719). Quoted by David F. N. (1962), in *Games, Gods, and Gambling*. London: Griffin.

CHAPTER XVII

MATHEMATICAL PROBABILITY IN RESEARCH

MODERN science owes a not inconsiderable debt to the gaming habits of the seventeenth-century French nobility.

The beginning of the mathematical theory of probability was with a question put by one of these inveterate gamblers to Pascal, who sent it on to, and discussed it with, Fermat, one of the most brilliant of a brilliant band of seventeenth-century mathematicians. It was an old puzzle. Two equally skilful gamblers have put up a stake, to go to the one who first wins so many points: when the first has scored p points and the second q, they have to break off. How should the stakes be divided? The solution aroused enormous interest, both at the gaming tables and among contemporary mathematicians. The gamblers wondered if the mathematicians could give them a rule by which they could always win: the mathematicians, painfully aware that they might be useful to a set of useless people, were fascinated by the abstract notion of probability. And so, from dice and cards was conceived the calculus of probability, which Pascal and Fermat are credited with founding.

Everyone agrees that mathematical probability is an important concept in modern science. Its usefulness is not impaired by the unexpected fact that no one knows what it means.

The distinction between 'credibility' and 'probability' is clear and well understood—the one concerns degrees of belief or doubt, the other a fact about a certain class of events. It is the second, *mathematical probability*, whose interpretation is in doubt.

No mathematician disputes the collection of formulae, described as the *calculus of probability*, which can be deduced from some simple axioms. Bertrand Russell has pointed out that it often happens that we know a mathematical formula to be true, but can find no meaning, or a number of different meanings, for some of the symbols. Given two or more interpretations satisfying the axioms, the choice between them is one of convenience: one is not correct and another incorrect.

By far the most convenient and the most simple interpretation of 'probability' is what Russell has called the *finite frequency theory*. The following definition may be stated:—

If α is a finite class of n members, m of which are also members of a class β, the probability that any α *taken at random* is a β is m/n.

For example, if there are 52 cards in an ordinary pack, of which 4 are aces, the probability that any card drawn at random will be an ace is 4/52. This is not to say that probability is a proportion: the proportion of $\alpha\beta$'s to α's. An essential part of the definition is *taken at random*.

An α is taken at random from a collection of α's when the method of taking it ensures that all the individuals in the collection will in the long run be taken with equal frequency. The relative frequency, therefore, with which any α that is also a β will be selected is m times in every n

trials, in the long run. The relative frequency with which any card taken at random will be an ace is, in the long run, 4 times out of every 52 trials.

'Probability' is being used for the relative frequency of the occurrence of a random event. 'Mathematical probability' has no meaning outside the domain of random events.

You may ask: 'How do we take things at random?' You will be told, by those who have to do it in planning experiments, that the best method, completely eliminating any possible bias in the selection, is to use a table of random numbers. You could, for instance, give every α a number from 1 to n; then read out from the list of random numbers, putting an α into the sample as its number turned up, until you have a sample of the size desired. A table of random numbers can be prepared by some procedure analogous to tossing pennies, rolling dice, shuffling and dealing cards, spinning a roulette wheel, and other entertaining activities. Any such procedure results in the selection of a thing from a set of things of the same sort, any one of which will be selected in the long run with the same frequency as any other.

When we say of a thing, or collection of things, that it has been taken at random, we mean that only 'chance' has operated in its selection. If one of the principal requisites of a scientific language is that every name must have an unmistakable meaning, the name 'chance' is a disgrace to scientific nomenclature. I have tried to present the idea of 'probability', and of 'randomness', without using it. Nevertheless, it might be a useful word if its meaning in any context were unmistakable.

Anyone who thinks about 'chance' will soon be completely bewildered if he does not first disentangle its several popular meanings.

'I met Smith *by chance* this morning': the meaning here is simply that I did not meet him *by design*. We often use 'chance' as the antithesis of 'design': 'Do you seriously believe that life came to the earth by chance?'

Again, it is often clear that 'by chance' is used with the meaning of 'by coincidence': 'By chance Shakespeare died on his birthday.'

The 'chance' with which we are concerned in mathematical probability is the gambler's 'chance'. To everyone in the heyday of gambling in classical Roman times it meant the goddess controlling the fall of the dice; and there are people today who seem to have much the same idea.

Cicero tried to disillusion his compatriots but seems to have substituted a notion just as erroneous—that chance events are uncaused. This also, I think, is still a common belief. It must be firmly rejected: events that happen by chance must be determined by something or other.

In saying that an event happened by chance we are not saying anything about our ignorance or uncertainty. We are saying something positive about the nature of the event and how it happened, though as a rule we are quite vague about what it is we are saying.

'A very small cause which escapes our notice determines a considerable effect that we cannot fail to see, and then we say that the effect is due to chance.' Poincaré suggested that at times this is all we mean. As an example, think of a cone precariously balanced on its tip. It will fall, but what makes it fall is a force so slight that it escapes our notice, and we say it falls by chance.

The cone will fall, but exactly in what direction is unpredictable. It is determined by a very slight force sufficient to shift the cone's centre of gravity. If, on successive occasions, this force is subject to regular variations in direction, it will have the effect that any way of the cone's falling will occur, in the long run, with the same frequency as any other. I do not know if a balanced cone does fall, in successive trials, in every direction with the same frequency. But if it did, the cause would lie in the regularity of slight variations in the force that upsets its balance. We may accept that a penny falls heads and tails in the long run with the same frequency; or that one of the six faces of a true die turns up after a fair roll with the same frequency as any other face. And this would seem to imply small but regular variations in the forces that determine how a die comes to rest, or a penny falls.

Such very small causes of a regularly varying character generate a special class of events, such that, though an individual event is unpredictable, all occur with the same frequency—the class of random events, individually unpredictable, but equiprobable with others of their set.

It is not open to anyone to say, of a word of common speech, how it should be used; all he is competent to do is to say how it is used. It would seem that one of the uses of 'chance', and perhaps the most important, is to denote the sum of a number of small causes which generate the class of random, or equiprobable, events.

Probability theory is now an indispensable tool in scientific research. It has provided the key to a number of biological problems, usually by showing how we may calculate the probability of some event; and it is because 'chance has rules which can be known' that such calculations are possible.

The mathematical theorems are deduced from 6 axioms. The first 4 assert that there is only one value of a probability; that its possible values are all the real numbers from 0 to 1; that 1 denotes certainty and 0 impossibility.

The fifth axiom, called the *conjunctive axiom*, relates to the joint occurrence of several independent random events whose separate probabilities are known. The probability of the compound event is *the product of their separate probabilities*. If, for example, three dice are thrown together, the probability that all will turn up 'six' is $1/6 \times 1/6 \times 1/6 = 1/216$.

The sixth axiom, called the *disjunctive axiom*, relates to several alternative random events which cannot occur together: in the terms of the definition, it gives the probability that any α taken at random is a β, or a γ, or ..., if the classes are mutually exclusive. The probability required is *the sum of the separate probabilities*. The probability of getting either a 'one' or a 'six' from a single throw of a die is $1/6+1/6 = 1/3$.

Calculation of some random events is simple, in that it needs only a knowledge of basic concepts, and either a direct application of the conjunctive or disjunctive axiom, or a calculation of the n and m of the definition.

To give a few examples:—

If a spin of a coin comes down heads 5 times running, what is the probability that it will come down heads on the next spin? The probability is not affected by what has gone before: it remains 1/2.

If two cards are drawn at random from a pack, what is the probability that both are red? The probability that the first card is red is 26/52; the probability that the second card is red is 25/51: therefore the probability that both are red is $1/2 \times 25/51$. Drawing the first card has altered the data: for the second draw there are 51 cards, of which 25 are red.

In throwing two dice, what is the probability that the numbers will add up to 8? The *n* of the definition is 36, for the dice may fall in 36 ways; the *m* is 5, for in 5 of these ways the numbers add up to 8. The probability required is therefore 5/36.

It may well be that in asking for the probability of some event we are not indifferent to the answer. Two normal parents have produced a child with some hereditary defect—say, albinism, or amaurotic idiocy—and ask, what is the probability that the next child will be so afflicted? The question does not differ in kind from the first example, about the spin of a coin. If they can be told, as in these instances they could be, that the probability is 1/4, it will be wise to make sure that in their anxiety they do not read too much into the information. They may think that they can feel quite happy about the next 3 children. In fact, a particular random event is quite unpredictable. It is impossible to predict whether a child of parents such as these will or will not have the defect. In saying that the probability of albinism is 1/4 we are making a prediction about a class—the class of albino children of such parents: that they will, in a very large number of such families, be found with a frequency, relative to all the children, of 1 in 4.*

An insurance company will not predict how long you will live: it is not even interested. The predictions it makes are about the class of lives to which you belong—healthy men, or women, of a given age.

If the parents think: 'Well, at least it is probable that the next child will be healthy', they are unconsciously leaving 'mathematical probability' for 'credibility'. They are validly inferring from your information: 'The next child will be healthy has some degree of credibility.'

One great use of probability theory in biology is in the science of genetics. The simple basic concepts are its foundation, for genetic hypotheses and their consequences involve random behaviour.

Let us look again at one of Mendel's experiments. He crossed a pure-breeding round-seeded garden pea plant with a wrinkled-seeded plant, and the following year grew the hybrids. Denoting the dominant factor for roundness of seed by 'A', and the recessive factor for the wrinkled character by 'a', the pair of factors in the F_1 hybrid is 'Aa'.

When these hybrids make their egg cells and pollen cells, the factors separate: two kinds of eggs (A) and (a) are formed *in equal numbers*;

* That a Mendelian ratio will be found in a small family is a common mistake, which, with others, appears in the well-known limerick:—

There was a young woman called Starkie,
Who had an affair with a darkie.
The result of her sins
Was quadruplets, not twins—
One black, and one white, and two khaki.

and two kinds of pollen cells (A) and (a) *in equal numbers.* Mendel allowed the hybrids to self-pollinate. His theory required that any egg cell has an *equal chance* of being fertilized by any pollen cell. Still fundamental in the theory of the gene is the *random union* of gametes at fertilization. Fertilization is a random event.

A very helpful model of what we are talking about in discussing random events is the urn containing black and white balls. It was first suggested by Laplace (1747–1827), one of the mathematicians who contributed to the calculus of probability; and he had borrowed it from the technique of State lotteries, which was a periodic withdrawal at random of a numbered ticket from an urn full of numbered tickets.

Imagine a very large urn, marked *m*, of black and white balls in equal numbers, to represent the male gametes: black$_m$ for (A), and white$_m$ for (a); and a second very large urn marked *f*, of black and white balls in equal numbers to represent the female gametes, black$_f$ for (A), white$_f$ for (a). Now, time after time, remove one ball from one urn and one from the other, at random. As two are not taken from one urn, only 4 pairs are possible: black$_m$, black$_f$; black$_m$, white$_f$; white$_m$, black$_f$; white$_m$, white$_f$. The mathematical theory predicts that, of many such pairs, the relative frequency of 2 black balls should be 1/4, of a black and a white 1/2, and of 2 whites 1/4.

Mendel's hypothesis was that the gametes should behave precisely as the black and white balls in this model: so that one-quarter of the seeds formed by the self-pollination of F_1 hybrids should have the genetic constitution AA, one-half Aa, and one-quarter aa. 'A' being dominant, three-quarters of the seeds should be round and one-quarter wrinkled. Not that the proportion 3 : 1 should be found in every 4 seeds, but that in a very large number of such samples the mean number of round seeds should be 3.

What a hypothesis of this kind predicts is that some proportion will be found *in the long run.* In the short run anything may happen. In the long run spinning a coin time after time will result in heads turning up with the same frequency as tails; in the short run we may easily get a long succession of tails. In genetic research enormous samples are desirable to confirm or refute a prediction of this sort. This is one reason for the great difficulty of research in human genetics.

Mendel looked at 7324 seeds, obtained by the self-pollination of 253 hybrid round-seeded plants: of these, 5474 were round and 1850 wrinkled. It was a close approximation to the proportion required by his theory.

It would be impossible now to do research in genetics, especially in human genetics, without at least an elementary knowledge of mathematical probability.

Requests for the probability of some event are not all answered so easily as those so far given as examples. Some, though of the same elementary kind, need the use of a mathematical formula that can be deduced from the axioms.

I refer to the type of question: If p is the probability that an event will happen in a single trial, and q the probability that it will not happen, what is the probability, P, that it will happen r times in n trials?

At first sight it may appear trivial, but it has proved every bit as fruitful as the famous 'points question' of the seventeenth-century gamblers. Resorting to the urn of Laplace, let it be filled with black and white balls in equal proportions and uniformly mixed. If a ball is withdrawn at random the probability, p, that it will be black is 1/2; the probability q $(= 1-p)$ that it will be white is 1/2. Remove, say, 4 balls at random—one at a time if you like; it comes to the same thing—what is the probability that, say, 3 will be black?

The answer was found in the late seventeenth century by Jacques Bernoulli, one of a family which produced nine illustrious mathematicians. The probability of drawing a black ball $n, n-1, n-2, \ldots, 0$ times in n trials is given by the successive terms of the expansion of $(p+q)^n$.

We simply look for the appropriate term. For drawing 3 black balls in a sample of 4, the probability is the appropriate term in the expansion of

$$(p+q)^4 = p^4+4p^3 q+6p^2 q^2 +4pq^3+q^4.$$

As black is represented by p, the appropriate term is the term containing p^3, namely, $4p^3 q = 4\times(1/2)^3\times 1/2 = 4/16$.

As the sample size increases the labour of multiplying out $(p+q)^n$ becomes enormous, but it is quite unnecessary. The *binomial theorem* shows that the coefficients of the successive terms of the expansion of $(p+q)^n$ are ${}^nC_r$, and the powers of p and q are $p^r q^{n-r}$, giving r the successive values $n, n-1, n-2, \ldots, 0$.

The term giving the probability required by the general question, that an event will happen r times in n trials, is the term containing p^r, which is

$${}^nC_r p^r q^{n-r}.$$

${}^nC_r$ is the symbol for the number of ways in which r things can be chosen from a collection of n things; it is:—

$$\frac{n!}{r!(n-r)!},$$

which may be written:—

$$\frac{n(n-1)(n-2)\ldots(n-r+1)}{r!},$$

The combined use of this formula and the disjunctive axiom is needed to answer this very common type of question: What is the probability that, in a family of 6 children, at least 3 will be boys? It is the sum of the separate probabilities that 6, 5, 4, 3 children will be boys; and these are obtained, of course, by putting the successive numbers for r in the formula.

Perhaps the most fruitful of all applications of the mathematical theory of probability in research has been the solution of problems arising from the natural differences between random samples of the same thing.

We would all expect any sample of a bottle of wine to be much the same as any other. Samples of any perfectly homogeneous substance show no variation, but things are very different with samples of a collection of discrete things.

If anyone is vague about the degree to which such samples can vary, he will find that knowledge can be made precise by an application of the Bernoullian law. Going back once again to the urn with its black and white balls in equal proportions, remove a ball at random 10 times in succession. You have a random sample of 10: it may contain 5 black and 5 white balls, but more likely not. The formula gives the relative frequency with which this proportion will be found in a very large number of samples ($n = 10, r = 5, p = \frac{1}{2} = q$):—

$$P = \frac{10\times9\times8\times7\times6}{5\times4\times3\times2\times1} \times(\tfrac{1}{2})^{10} = 252\times\frac{1}{1024}.$$

The relative frequency, or probability, required is 252 times in 1024 samples, or 0·246: approximately 1/4.

The frequencies with which 0, 1, 2, ..., 10 black balls should appear in the samples, as predicted by the mathematical theory, are shown in the second column of *Table II*. It will be seen that, while, in a large number of samples of 10, about a quarter will show equal numbers of black and white, almost as many will contain 6 black and 4 white, or 4 black and 6 white. Three black and 7 white, or 7 black and 3 white, are seen much less often, and the frequency tails off rapidly as the disproportion increases. Ten black and 0 white, or 0 black and 10 white, will be found slightly less than once in every thousand samples in the very long run.

Given the axioms the formulae of the mathematical theory are without any doubt correct. But when we use a formula as a model of what happens in the random behaviour of physical things, certainty disappears. That random events follow mathematical rules is a hypothesis and needs verification like any other.

That spinning a coin will turn up heads and tails in the very long run with the same frequency is a hypothesis. The very long run means *ad infinitum*. I have not heard of anyone who has had the stamina to spin pennies more than a few thousand times. The eighteenth-century naturalist Buffon astutely got a young child to do it for him, and he obtained 1992 tails to 2048 heads. Augustus de Morgan, the English mathematician and logician, persuaded a young pupil of his to do the same, and he recorded 2044 tails to 2048 heads. Stanley Jevons, throwing up 10 coins at a time, recorded: 'The whole number of single throws of coins amounted to 10×2048, or 20,480 in all, one half of which, or 10,240 should theoretically give heads. The total number of heads obtained was actually 10,353. . . . The coincidence with theory is pretty close, but considering the large number of throws there is some reason to suspect a tendency in favour of heads.'

Stanley Jevons also compared the theoretical frequencies of an event happening 10, 9, 8, 7, 6, 5, 4, 3, 2, 1, 0 times in 10 trials, given in *Table II*, with his own observations. Although black and white balls are an excellent model of mutually exclusive characters, the experiment is much more easily carried out by spinning coins. It does not matter in the least what the two mutually exclusive characters are, or what has them. Ten successive spins of a penny will provide a random sample of 10 from an unlimited population of heads and tails in equal proportions. Jevons says in *The Principles of Science*: 'Taking a handful of ten coins, usually shillings, I threw them up time after time, and registered the number of heads which appeared each time. Now the probability of obtaining 10, 9, 8, 7, etc. heads is proportional to the number of combinations of 10, 9, 8, 7, etc. things out of 10 things. . . . I made altogether 2048 throws, in two sets of 1024 throws each.' The numbers he recorded are reproduced in *Table II.*

Table II.—THE THEORETICAL FREQUENCIES OF AN EVENT COMPARED WITH ACTUAL OBSERVATIONS. (*After Jevons.*)

r Heads	$^{n}C_{r}$	Probabilities	Jevons		
			First Series	Second Series	Average
10	1	0·0009	3	1	2
9	10	0·0097	12	23	17½
8	45	0·0439	57	73	65
7	120	0·1171	129	123	126
6	210	0·2051	181	190	185½
5	252	0·2461	257	232	244½
4	210	0·2051	201	197	199
3	120	0·1171	111	119	115
2	45	0·0439	52	50	51
1	10	0·0097	21	15	18
0	1	0·0009	0	1	½
Totals	1024	0·9995	1024	1024	1024

We may now look at an example of a problem arising from the inconstancy of samples, which is solved by the calculation of a probability, a calculation arrived at by an application of the Bernoullian law combined with the classical law of errors mentioned on p. 75.

It is one of many problems in genetics, where we have to say whether observed proportions are in accordance with those predicted by the hypothesis. It is a question of *goodness of fit*. Consider once more Mendel's results in the experiment described earlier. In a sample of 7324 seeds he observed 74·7 per cent round and 25·3 per cent wrinkled; his theory predicted 75 per cent round and 25 per cent wrinkled. This degree of approximation seems good enough, and we would probably pass 74 per cent or 76 per cent. But would we accept 77 per cent, or would we say that the hypothesis had predicted what was not a fact, and therefore should be rejected?

What we want is to be able to deal with a critic who says: 'This is too big to be a chance difference: it is all very well for you to say that your results agree with what theory predicts *within the limits of experimental error*; the only source of error in your experiment is the inconstancy of samples, and this does not account for so large a discrepancy.' How do we fix the limits of experimental error?

The problem involves comparison of a sample with its known, though hypothetical, population.

We apply a test, the principle of which is to find by a calculation if the probability, P, of a chance discrepancy of this magnitude is greater or less than 0·05. If it is greater it means that a discrepancy of this size will occur by chance more often than once in 20 trials, and we claim that chance explains it. If it is less than 0·05 it means that the difference was so big that it would occur by chance less than once in 20 trials, and we reluctantly admit that the critic may be right.

As a criterion the level $P = 0{\cdot}05$ is a convention; if we wish to be stricter we can choose $P = 0{\cdot}01$, once in 100 trials.

The probability, P, of any observed discrepancy having occurred by chance is the key to interpretation. In very brief outline, we are able to determine it by reasoning in this way.

A large enough number of random samples of 10, of a population of two mutually exclusive characters in equal proportions, will show any specified number of one of these characters with the relative frequency set out in the second column of *Table II*. This is a rather remarkable series of numbers. It will be noticed that they are distributed symmetrically about the most common frequency—that of 5 of the character, or in general $n/2$. This distribution is referred to as a 'binomial' or 'Bernoullian' distribution.

If these frequencies are plotted as a graph, they fall on a curve such as that described on p.75 as the 'curve of errors' or the 'normal distribution curve'.

The reasoning is based on the mathematical properties of the normal curve.

The predicted proportion may be considered the known proportion in a hypothetical population, of which we propose to observe a sample. When we know the population proportion we can calculate its variability in a set of samples. The amount of variability is measured by the 'standard deviation', or *standard error*, and is easily calculated from a simple formula.

If p is the proportion of the character in question, and $q = 1 - p$ that of the objects that do not possess it, then p will vary in a set of samples of size n with a standard error $\sqrt{(pq/n)}$. If we prefer to speak of percentages, p denotes the percentage of one, $100 - p$ that of the other.

The properties of the normal curve enable us to assert that if the difference between the expected proportion and the observed proportion is *more than twice the standard error*, it is unlikely to be a chance difference: for the probability of a difference as large or larger than this is 0·05 or less. But if the difference is less than twice the standard error it could easily be due to chance: for the probability of a chance difference smaller than twice the standard error exceeds 0·05.

This reasoning has assumed that in the population the mutually exclusive characters are in equal proportions: $p = 0{\cdot}5 = q$. When the proportions are unequal, as they usually are, the curve becomes asymmetrical. For samples of 20 or more it is, however, only moderately skew, unless the disproportion is something of the order of 10 : 1. The test is only approximate, but is usually applicable.

It could be applied to Mendel's results as follows:—

The proportion required by theory is 75 per cent round seeds and 25 per cent wrinkled seeds. The standard error of a sample of 7344 is $\sqrt{(75 \times 25)/7344} = 0{\cdot}5$ per cent. We can therefore say that a difference between the expected and observed values of less than 1 per cent will occur with a probability of more than 0·05, and should therefore be accepted as a chance difference. Mendel's observed difference was in fact 0·3 per cent.

Standard error tests are not often used.

For the interpretation of all experimental results that could be due to chance, so-called 'tests of significance' have been perfected since the beginning of the present century. They became very popular in the 1930's, and now in some kinds of research are *de rigueur*.

The key which they provide is the same; the probability of a 'chance' explanation, whether it is above or below a level accepted by convention, usually 0·05. But the mathematics which gives us the key has become complicated. The test itself is applied by rule of thumb.

The χ^2 (chi-square) test was devised by Karl Pearson in 1900 and perfected by R. A. Fisher in 1922. It is used for goodness of fit, when the experiment has been an observation of the proportion of more than two classes. Mendel's experiment of crossing a garden pea with round yellow seeds with a variety with wrinkled green seeds, described on p. 46,

Table III.—THE X^2 TEST APPLIED TO ONE OF MENDEL'S OBSERVATIONS

CLASS	o OBSERVED FREQUENCY	e EXPECTED FREQUENCY	$o-e$ DEVIATION	$\frac{(o-e)^2}{e}$
Round and yellow	315	312·3	2·7	0·023
Round and green	108	104	4	0·154
Wrinkled and yellow	101	104	−3	0·154
Wrinkled and green	31	34·7	−3·7	0·394
Total	555	555	0	$\chi^2 = 0{\cdot}725$

is an example. His theory predicted that the F_2 seeds should be in the proportion: 9 round and yellow, 3 round and green, 3 wrinkled and yellow, 1 wrinkled and green. He obtained in his experiment 555 seeds, and his number for each class is shown in *Table III* in the column headed 'o'. The numbers under the column 'e' are those expected as a consequence of his theory.

The test is seen in action in the last two columns of the table. The value of χ^2 is the sum of $\{(o-e)^2/e\}$ for all classes: in this instance 0·725. By referring to a table of χ^2, in which the appropriate probabilities have been worked out for us, we can find approximately the probability of the chance occurrence of any χ^2 value. This naturally diminishes as χ^2 increases: for the value of χ^2 increases with the deviation of observed from expected frequencies.

Table IV is extracted from R. A. Fisher's much larger χ^2 table. It is in rows of χ^2 values for various numbers of 'degrees of freedom'; in problems of this sort this number is one less than the number of classes. In the row for 3 degrees of freedom we find that a value of 0·725 falls between the values of χ^2 corresponding to the probabilities 0·95 and 0·70, and so we feel confident that the result agrees 'within the limits of experimental error' with what was predicted.

Table IV.—TABLE OF χ^2

DEGREES OF FREEDOM	$P=0·99$	$P=0·95$	$P=0·70$	$P=0·50$	$P=0·30$	$P=0·05$	$P=0·01$
1	0·00016	0·00393	0·148	0·455	1·074	3·841	6·635
2	0·0201	0·103	0·713	1·386	2·408	5·991	9·210
3	0·115	0·352	1·424	2·366	3·665	7·815	11·341
4	0·297	0·711	2·195	3·357	4·878	9·488	13·277
5	0·554	1·145	3·000	4·351	6·064	11·070	15·086
6	0·872	1·635	3·828	5·348	7·231	12·592	16·812
7	1·239	2·167	4·671	6·346	8·383	14·067	18·475
8	1·646	2·733	5·527	7·344	9·524	15·507	20·090
9	2·088	3·325	6·393	8·343	10·656	16·919	21·666
10	2·558	3·940	7·267	9·342	11·781	18·307	23·209

All biology students are now taught the use of these tests. Most of them must blindly accept the authority of the mathematicians that they will determine the probability required with sufficient accuracy. Even if they are willing and able to follow the long and complex underlying algebraic argument, they will not find it set out in any of the more elementary textbooks. But in use the tests are very simple and straightforward. There is no need to agonize over what a 'degree of freedom' may be. There is no need to have the least idea of why the test works: the result will be the same.

At the end of Chapter XVI an important question was left unanswered. If, in the two groups of the experiment there described, we find some numerical difference, how can we refute a suggestion that the variability of samples will account for it? A critic will look at such results and assert that they are in accordance with what he calls a *null hypothesis*—that they show, not a real difference, but one due to chance.

The modern therapeutic trial is as good an illustration as any of this kind of experiment, and of the use of statistical tests to refute the null hypothesis.

Consider the difficulty of deciding whether a reputed therapeutic agent really does affect the course of a disease. We all know many things that do this: with only few exceptions, the course of any disease is affected by such things as age, sex, occupation, general nutrition, anxiety, genetic constitution, and many others. If a proportion of treated patients improve, how can we know that the treatment has been responsible for it?

It is not so very long ago that most physicians were content with 'feeling certain' that this or that treatment was effective, or that one was better than another. Instances of treatment once thought to be effective and now completely given up are innumerable. Consider the vaccine treatment of rheumatoid arthritis so widely used in the 1920's. There was no evidence that rheumatoid arthritis was an infective disease, and no theoretical reason for expecting vaccine treatment to help, even it it were. But clinical pathologists made innumerable autogenous vaccines from Streptococci in the nasopharyngeal flora, in tooth sockets, in the colon, and elsewhere; and stock vaccines were mass-produced to avoid these laborious preliminaries. Physicians were convinced of their beneficial effect because they saw patients improve during their administration. They forgot, or did not know, that many sufferers from rheumatoid arthritis improve if left alone.

Nowadays new remedies are first submitted to a properly planned therapeutic trial before being generally used. The old remedies continue. The requirement is that we shall observe two groups of patients as nearly as possible alike in those variable things that may affect the outcome of the disease, and that we treat both groups in exactly the same way and under the same conditions, with one exception: in one group we use also the therapeutic agent to be tested.

There is one apparent exception to the need for a second group not treated by the medicament under trial: this is when, after the administration of a new drug, a patient recovers from some hitherto uniformly fatal disease. Before 1935 or 1936 recovery from streptococcal meningitis was excessively rare: the illness was universally regarded as hopeless, with a mortality of almost 100 per cent. Then, with the introduction of sulphanilamide and related sulphonamide drugs, recoveries began to be reported in the journals, one or two every few weeks. The whole of past experience was adequate control.

The detailed planning of a therapeutic trial is difficult and need not concern us here. Usually it is desirable to restrict the trial to a particular type of the disease, rather than attempt the study of a sample representative of every type. In the now celebrated trial of *streptomycin* in pulmonary tuberculosis* it was decided to sample the population: 'Acute progressive bilateral pulmonary tuberculosis of presumably recent origin, bacteriologically proved, unsuitable for collapse therapy, age group 15–30.' This deliberate choice ensured that all patients had a similar type of disease, and naturally the results applied only to this type.

An essential part of the plan is the random allocation of treatment to one or the other of a pair. Every patient must have an equal chance

---

* Medical Research Council (1948), *Br. med. J.*, **2**, 769.

of falling into the control or the trial group. In the streptomycin trial this was secured by the use of a table of random numbers.

There is, however, one variable that may affect the result of the test that must be dealt with in another way—this is the psychological effect of treatment. If patients in the trial group know that they are being given a new remedy likely to help, and those in the control group know that they are having only the usual and not very successful treatment, part of the difference found may be due to optimism acting unequally in the two groups.

In the streptomycin trial the patients under test were not told that they were having a special treatment, nor those in the control group that they were controlling anything. It is more usual to give patients in the control group a dummy treatment indistinguishable from the treatment under trial; it may be an inert tablet looking and tasting the same. If all patients are told that some are having dummy tablets, optimism will fade and have no bearing on the result. Alternatively, all can be told that the tablet they are having is under trial; then optimism will affect members of both groups equally.

At the end of any allotted period of trial there will be numerical results to contrast in the two groups: it may be the numbers of those who have died and those who have recovered, or those who have completely recovered, improved, remained unchanged, or become worse. There are many possible sub-divisions.

Consider this observation selected from the results of the streptomycin trial, in which 52 patients in the control group were treated by bed-rest, and 55 patients in the trial group were treated by bed-rest and streptomycin: 14 (27 per cent) of the control patients and 4 (7 per cent) of the streptomycin patients died before the end of 6 months.

The *null hypothesis* asserts that streptomycin was inert, and a chance difference was recorded. Two random samples were taken, the critic will explain, one of 52 and the other of 55 patients, from sufferers from a particular form of pulmonary tuberculosis; and both were treated in precisely the same way, assuming streptomycin to be inert. The difference is simply the difference we might expect from the variability of samples.

The χ^2 test gives the actual probability of such a chance difference. It is applied as shown in *Table V*.

Table V.—THE χ^2 TEST APPLIED TO AN OBSERVATION IN THE STREPTOMYCIN TRIAL

	STREPTOMYCIN GROUP			CONTROL GROUP		
	o	e	$o-e$	o	e	$o-e$
Died	4	9·25	−5·25	14	8·75	+5·25
Survived	51	45·75	+5·25	38	43·25	−5·25
Totals	55	55	0	52	52	0

We do not know the proportion of those dying and surviving in the tuberculous population, of which, if the null hypothesis is true, these are samples. The best estimate is from the total figures: 18/107 dying and 89/107 surviving. From this estimate the expected values in each class are calculated:—

$$\chi^2 = (5{\cdot}25)^2 \times (1/9{\cdot}25 + 1/45{\cdot}75 + 1/8{\cdot}75 + 1/43{\cdot}25) = 7{\cdot}358.$$

Referring to *Table IV* in the row 1 degree of freedom, this value is greater than that corresponding to a *P* of 0·01. The probability of finding a difference as big as this in two samples purely from chance is therefore less than 0·01: it would be found in a pair of samples less than once in 100 pairs. The result is said to be 'significant'.

The null hypothesis in this instance is considered so improbable that we feel confident in rejecting it.

When the contrast to be made is between two sets of measurements different tests are needed. The principle is the same: they are devices to reveal the probability of a null hypothesis being true.

Facility in the use of these tests, which now seems universal, is perhaps not matched by a clear grasp of their limitations.

'Significant' is more important in sound than in meaning. All it means is that the result of the experiment is unlikely to be due to the inconstancy of samples, or, if you prefer, to chance. However badly planned your experiment the result may still be 'significant'.

The purpose of every experiment is to determine whether observation confirms what is predicted by the theory whose verification we are undertaking. No statistical test has any bearing on this theory, except that it removes, or renders improbable, an alternative hypothesis. This is important, of course, but no more so than any other experimental control.

When the results of a therapeutic trial are said to be significant, anyone uninstructed in its special meaning in this context would naturally understand that the therapeutic agent had been found effective, whereas the credibility of this conclusion depends on quite other considerations.

Research workers themselves understand this quite well, but the uninitiated who read their reports do not.

There is another risk which tends to be ignored even by some investigators. It is to forget that what is established by a statistical test is a probability. It is not possible to establish conclusively that the goddess either has or has not interfered.

When a difference between the treated and the control group is so small that the probability of its having occurred by chance is greater than 0·05, the investigator may nevertheless be wrong in accepting it as a chance difference. All he can say is that more often than once in 20 trials it would occur by chance: he could give you the exact frequency if he knew *P* exactly—once in 10 trials if it were 0·1; 3 times in every 4 trials if it were 0·75; and so on. But never can he say that it *must* have occurred by chance.

If, on the other hand, the test shows that the probability of his observed difference having occurred by chance is less than 0·05, and he thinks this permits him to exclude chance as the explanation, he will be wrong perhaps once in 20 times. If the probability works out at less than 0·01 he would be wrong in this way once in 100 times.

Cicero knew well enough that improbable things happen: 'Every man who plays often will at some time or other make a Venus-cast: now and then indeed he will make it twice, or even thrice in succession. Are we going to be so feeble-minded then as to aver that such a thing happened by the personal intervention of Venus rather than by pure luck?'*

It may be risky to forget that these tests are only approximate methods, and that what they approximately give is a probability, which we then make the basis of an inference about the credibility of our theory.

* CICERO, *De Divinatione*. Quoted by DAVID, F. N. (1962), in *Games, Gods, and Gambling*. London: Griffin.

CHAPTER XVIII

THE ART OF SOPHISTRY

ATHENS in the fifth and fourth centuries B.C. attracted teachers from all parts of the Hellenic world. They were called 'sophists'. They taught for money, which was not quite the thing to do in Athens, but in spite of this it was at first an honourable name.

There were some among them who taught what they called 'rhetoric'. They would teach you how to plead in the law courts, a very useful accomplishment, for at that time every citizen pleaded his own cause. They would indeed teach you how to argue any debatable point, for all Athenians loved talk and controversy. They would teach you how, if you had no case you could make it appear as if you had one, by the use of arguments persuasive and plausible even though invalid. This highly desirable knowledge could be yours for a consideration.

And so, although most of the teachers of ancient Greece were as intellectually honest as our own professors, and many, like Socrates, who never accepted a fee, were men of the highest reputation, in time 'sophist' came to mean a quibbling fallacious reasoner.

Sophistry still means something to which we would not stoop ourselves, though it is obviously as popular as ever with other people.

Aristotle showed how reason could reveal truth: the professors of rhetoric how crooked thinking could ensure success.

But success is not always contemptible, and not everyone makes a god of truth. You may say that desirable ends do not justify any means for their attainment, and that to stoop to sophistry in the pursuit of success, be it in the achievement of no matter how laudable an end, cannot be justified.

I could point to some very odd things that have been done in the name of truth. I could tell, for instance, of how one of the kindest and gentlest of men bored holes in the skulls of dogs, so that he could then inject saliva from some rabid animal over the surface of the brain, to see if they developed rabies. And how tens of thousands of men and women and children have blessed his name for doing it, and discovering the nature and prevention of the most terrible virus disease known to man.

But if I did, in this context, it would be an instance of that fallacious argument which at the moment I want merely to expound and not to practise. *Ignoratio elenchi*, the medieval followers of Aristotle called it: arguing the wrong point. We are not discussing whether animal experiments are justified in the pursuit of truth about diseases; it is entirely irrelevant to the point at issue.

The point is whether sophistry is or is not objectionable.

Most readers of this book will be scientists, or at least will admire the scientific outlook. You would not be influenced for a moment by sophistry, however persuasive; you would ask for the evidence. But let me remind you that the beliefs of most people are based, not on evidence, but on testimony.

Suppose you feel it imperative that your fellow men should know the truth about something on which you are an authority, say something comparable in importance to the causal association of cigarette smoking and lung cancer.

You will not invite them to weigh the evidence. Few men are trained statisticians. Indeed the great majority of people cannot follow a closely reasoned argument; logical evidence is meaningless for them. You will, if what you want to persuade them of is not open to question, simply establish your qualifications to make the assertions you do.

If, however, your thesis is something that can be and is being questioned, your task is more complicated: convincing men that you are the authority to be accepted now involves discrediting those who contradict you, and making your thesis appear reasonable to those who cannot understand reason. You have to persuade people by methods other than a closely reasoned argument—methods that were taught to their pupils by the Greek sophists.

Is there a scientist who has never made use of them? Think of what some scientists said about Darwin, to discredit if they could his theory of descent. It was not only the theologians who resorted to misrepresentation, ridicule, and denunciation. The Cambridge professor of geology, for instance, wrote to him saying that if his doctrine were accepted, 'humanity would suffer a damage that might brutalize it, and sink the human race into a lower grade of degradation than any into which it has fallen since the written records tell us of its history'. There is no need to multiply instances of scientists trying persuasive argument which they know will not stand the test of rigorous logic.

The sophists of ancient Athens were teachers of 'rhetoric'. Much of what they taught can be found in the speeches of statesmen with the gift of persuasive oratory, or in the writings of journalists who can write with the pen of Voltaire. What is it then about sophistry that moves us to contempt?

Is it that these methods of persuasion are very often seasoned with logical fallacies: and we, who believe that reason is one of the angels guiding the path of man, dislike fallacies?

I cannot improve on Archbishop Whately's definition of a fallacy, more than a century ago: 'Any unsound mode of arguing, which appears to demand our conviction, and to be decisive of the question in hand, when in fairness it is not.' Now, if an argument to show the truth, as we know it, to men who cannot follow and cannot be expected to follow the evidence, appears to demand their conviction, is it always objectionable if in strict fairness it does not? Remember that a true belief is not falsified by a fallacious argument purporting to demonstrate its truth.

In one way and another I have had a good deal of experience of sophistry and its value in the inculcation of sound beliefs in those unused to the principles of rigorous logic. Like most doctors I remember with gratitude how much our teachers relied on *argumentum ad superiorem*. But rather more seriously, persuasive oratory or writing, even though using one or two arguments of doubtful logical validity, is, I suggest, rarely objectionable, and it often contains none—it is just an appeal to our hopes and fears, our sense of justice, and other emotions.

It has been said, and it may well be true, that 'in the answers to Parliamentary questions as much as in the writings of advertising copy, the ability to create an impression with words and phrases which are devoid of real content is reckoned by many of the professionals to be the hallmark of the real sophisticate'; nevertheless, it will be admitted that there are many occasions when methods of persuasion that are not closely reasoned arguments are desirable, and as the methods of a skilful sophist can be so described it follows that sophistry is at times desirable.

If the reader rejects this, I would ask him to consider that if all persuasion other than by reason is objectionable, then the preaching of all theologians is objectionable, for they would admit that, in our sense of the word, they have no evidence. They cannot give closely reasoned arguments for the beliefs they wish unbelievers to share, and therefore the rule is made untenable by this exception to it.

We may conclude that sophistry can be unobjectionable, if it be granted that the worth of great oratory or writing that has guided and inspired mankind has not been diminished by a few unsound arguments.

I shall give a brief account of the methods used by a skilful sophist. If the reader regards sophistry as a useful accomplishment, he may be helped by it, and if he thinks it a contemptible and dishonest trick he will be the better able to detect and expose it when used by his friends.

If I have persuaded you that sophistry is not necessarily objectionable, then you have been convinced by a string of sophisms.

I admitted to the first, in the hope of putting you off the scent. *Ignoratio elenchi*, literally 'ignorance of confutation', means 'arguing the wrong point'. It can be usefully introduced time and again in any lengthy polemic, and is a favourite technique for discrediting an opponent. The principle is to prove what he never denied, and to disprove what he never asserted.

Strictly, any attempt to persuade your audience to disbelieve your opponent, other than by proving the contradictory of what he says, comes under this heading. If, in this class of sophisms, we include irrelevant conclusions that bear no resemblance to the point at issue, there are many variants.

The familiar *tu quoque*, if you can make it stick, is always useful. After a little flattery, saying that, being scientists you would never be taken in by sophistry—flattery always helps—I tried one by asserting that scientists themselves at times make use of sophistry. Practise what you preach, I said: hoping you might think it had some bearing on the question. It has none.

Some of these variants of arguing the wrong point are so commonly used that Aristotle and the medieval scholastics gave them special names. *Argumentum ad hominem* means 'being a bit personal'. You prove something regrettable about your opponent's character, or merely abuse him. No better example of the use of this technique could be given than the prolonged emotional abuse and ridicule of Darwin evoked by the *Origin of Species*. Argument descended to vulgar derision. 'Are you descended from a monkey on your grandmother's side, or your grandfather's?' the Bishop of Oxford asked Huxley, who was defending Darwin's theory. More often than not abuse of your opponent is the instinctive emotional response of men who see some cherished belief in danger. Its deliberate use is not recommended.

After trying to enlist your confidence in me by saying how great was my experience, I pointed with approval to the technique of appealing to the emotions. And very persuasive it usually is, to pack as much emotion into your language as you can. Read again the criticisms of Darwin's theory of descent: good convincing stuff!

I then gave you an atrociously invalid syllogism: 'Some methods of persuasion not closely reasoned arguments (*M*) are desirable (*P*); the methods of a skilful sophist (*S*) are methods of persuasion not closely reasoned arguments (*M*): therefore, some sophistry (*S*) is desirable (*P*).

First, it is an instance of the fallacy of the undistributed middle term (*see* p. 20). In the symbolism of traditional logic its form is seen to be: Some *M* is *P*; all *S* is *M*: therefore some *S* is *P*—some Europeans are Frenchmen; all Germans are Europeans: therefore some Germans are Frenchmen. Exposed like this it would not deceive a child: bury it in a mass of verbiage, the recognized technique, and your audience may be too weak to detect it.

Again, the reader will probably accept the first assertion containing 'desirable' because he thinks it means: 'What men do in fact desire'—those who use such methods clearly do; but a persuasive sophist will make it appear that 'desirable' in the conclusion means: 'What should be desired.'

This shift of meaning, which makes two terms instead of one, is called the *fallacy of equivocation*. These fallacious arguments are not easy to invent; usually they are unintentional, deceiving the arguer as well as his audience.

Some names of common speech have a meaning that is single and clear. Such names are called *univocal*. Very many have two or more meanings: names called *equivocal* or *ambiguous*. Both are probably outnumbered by names whose meaning is vague. Confusion in discussion often arises from the ambiguity of names, and especially from the vagueness that borders on ambiguity. When a name is used in two senses in the same argument, and the change from one meaning to the other is such that it may well pass unnoticed, we have the fallacy of equivocation.

Names that have two or more completely different meanings, like 'church', 'foot', 'vice', 'mean', and so on, are not a source of this fallacy, for the change in meaning could deceive no one. To rank as a fallacy an argument must have some cogency. The fallacy of equivocation arises only with names that we have not realized are ambiguous, or with names so vague that a shift in meaning is hard to detect. To give an example: it has been pointed out that, in discussions of the Principle of Indeterminacy, there may be confusion from ambiguity of the word 'determined'; in one context it means 'measured', in another 'caused'.

The principle states that both the position and the momentum of an atomic particle cannot be *determined.*

It is usually wise, in discussing a difficult topic, to state the sense in which you are using a name that may appear ambiguous; though there is a risk of being accused of choosing a special meaning to suit your own purpose. Consider, for instance, the assertion of a celebrated scientist who should have known better: 'Reason has few discoveries to its credit.' 'Reason', which is the faculty of reasoning, should in this context mean 'those principles to be observed in the conduct of scientific inquiry'. If this is the meaning, the assertion is rubbish. But there are still many scientists, probably brought up on Aristotelian logic, who by 'reason' understand deductive inference. With this meaning, 'reason has few discoveries to its credit' is true and trivial: it refutes what no one has asserted since the Middle Ages.

After my faulty syllogism, which I am afraid will have deceived no one, I went on to the fallacy of *secundum quid*, claiming that the rule that only reasoned argument was unobjectionable was falsified by an obvious exception.

This is a very common and plausible form of fallacious refutation, whose full name is *fallacia a dicto simpliciter ad dictum secundum quid.* A general rule has been asserted without qualification. An instance that is manifestly false is then produced to refute it. But the so-called 'instance' ignores certain restrictions that for one reason or another were not mentioned, either for being so obvious or so rarely occurring: as if I tried to refute the usually accepted 'every man has a right to withhold his labour', by objecting that a surgeon may not. The unfairness lies in treating the rule as intended to apply without any qualification whatsoever, when I and everyone else knows it is not.

Probably the man who first said: 'The exception proves the rule' was thinking of a *secundum quid.* The exception appears to be an instance of the rule, but is not: its examination leaves the rule intact, but probably more precisely formulated.

After a recent trial the Judge said to one of the accused: 'If people do what they really believe to be right they cannot do much wrong.' In the correspondence columns of a national newspaper someone fell into a typical *secundum quid* by complaining that the generalization was false. Some people, he said, thought that corporal punishment was right, and euthanasia; and Hitler, he pointed out, thought it right to exterminate the Jews.

He was trying to demonstrate the falsity of an assertion, not as explicit as perhaps it should be, of what as a general rule is true, by pointing to a special case that is clearly untrue. But his special case ignored the implied conditions under which the rule is true. The Judge could have given all the restrictions beyond which the rule does not apply, but at the expense of being intolerably prolix. To his simple precept: 'If people do what they really believe to be right they cannot do much wrong' he could have added: 'provided they have been brought up in a civilized community and have average intelligence, provided their moral sense is not perverted by a psychopathic disorder', and so on. But he assumed that the girl to whom he spoke would take it as the simple guide it was intended to be.

Finally, I concluded by *begging the question*—always a good thing to do if you believe in sophistry and are hard pressed. An argument begs the question when the truth of the conclusion must be known before the premisses can be asserted.

One of the oldest criticisms of syllogistic reasoning is that it is a *petitio principii*; it begs the question. The point was first made about A.D. 200. It was said that we could not assert that 'all men are mortal' unless we had first noted the death of all men, including Socrates.

This is certainly just if the universal premiss is of the kind called 'enumerative'. Suppose I learn that a microscope has been stolen from a laboratory, and I can say at once that it is a Zeiss. Asked how I know this, I may say: 'Because all the microscopes in this laboratory were Zeiss.' The syllogism giving my conclusion is obvious, but it would be a trivial question-begging argument, for I could not assert that all microscopes were Zeiss unless I had examined each one, including the stolen one.

But the general propositions that we use in inference are not of this kind: they are, as has already been explained, universal judgements asserting that something is necessarily true of a class, and not a record that something has been found true of a finite and observable collection.

A commoner instance of 'begging the question' is to infer a proposition from the same proposition disguised by synonyms, for example, 'opium induces sleep because of its soporific properties'; 'every bodily and mental power is increased by good spirits because a general emotional satisfaction increases vitality'. Closely similar is a deduction from premisses asserting an analysis of the conclusion: 'All our decisions stem from the unconscious, because emotion, intuition, and instinct always have the last word.' The reason I gave for concluding that sophistry could be unobjectionable was just an analysis of this conclusion.

Petitio principii is, strictly, not a fallacy; it is merely trivial. The conclusion does follow from the premisses, for every proposition implies itself.

INDEX

Printed in Dunstable, United Kingdom

91230643R00070